Fostering Innovation for Agriculture 4.0

Miguel Angel Rapela

Fostering Innovation for Agriculture 4.0

A Comprehensive Plant Germplasm System

Miguel Angel Rapela ⓘ
Intellectual Property Centre
Austral University
Buenos Aires, Argentina

ISBN 978-3-030-32495-7   ISBN 978-3-030-32493-3   (eBook)
https://doi.org/10.1007/978-3-030-32493-3

This Springer imprint is published by the registered company Springer Nature Switzerland AG
The registered company address is: Gewerbestrasse 11, 6330 Cham, Switzerland

This book is dedicated to my wife Beatriz.

Preface

The innovation and scientific and technical development of modern plant varieties (which are novel, distinct, uniform, and stable), including the development of beneficial microorganisms, the access to and use of plant genetic resources, and the development of biotechnological and biosafety inventions, are regulated at the international, regional, and national levels. This regulation is in the form of many treaties, conventions, protocols, international agreements, and other regional and domestic rules. This complex set of rules has resulted in challenges to make global interpretations, due to overlapping, gaps, ambiguities, contradictions, and lack of consistency. The big picture is even more complex, as a series of scientific developments applied in plant breeding in general—especially in gene-editing techniques—have rendered these international regulatory frameworks obsolete.

Feeding and providing energy to the world requires doubling agricultural production between 2010 and 2050. Attaining this goal demands a yearly 2.4% growth rate in the main crops. A series of studies and analyses from different sources point to the fact that the productive growth rate of the main crops is in a critical point at half that value.

Indicators show that with the current scenario the world is facing an imminent and growing agricultural crisis in a perfect Malthusian model, in which Boserup's speculation that past a critical point new agriculture forms evolve and innovation is stimulated might be wrong.

The claim in this study is that facts are interrelated. The complex set of regulations about the development of modern plant varieties and beneficial microorganisms is the *cause* affecting the access to and use of genetic resources, as well as the research and development in genetics and plant breeding. And the *effect* is the standstill in productivity growth, that is, the innovation rate.

The attempts of methods, ideas, and proposals to tackle the regulatory hassle have not proven to be effective so far. To a large extent, this has been due to a lack of understanding of the productive models which have led to the development of "Agriculture 4.0" with features never seen before and which requires adopting a different regulatory paradigm in its broadest concept. The current "Agriculture 3.0" paradigm based on a set of separate regulatory tools does not seem to be the proper way to satisfy the demand in this new scenario.

Attempted solutions to these challenges based on game theory have also been unsuccessful. This work claims that this failure was due to resorting to models within a framework of a noncooperative, competitive, and mainly zero-sum game theory, without any consideration for the existing relationships between all players. Opposite to that, the proposal contained here is a solution based on applying a convex superadditive cooperative model of relationships among the three sectors or players—for this, it is necessary to design a system of balance and transfer of profits of participants.

Based on this, a "Comprehensive Plant Germplasm System" is introduced as a general proposal for all industries driving plant innovation, including heterogeneous plant varieties, microorganisms, biotechnological developments, genetic resources, and biosafety.

The system is a proposal of minimum contents for a binding international convention which would supersede all conventions, treaties, and other regional or domestic regulations covering native varieties and traditional developments, heterogeneous plant varieties, plant varieties, microorganisms, biotechnological inventions, plant-breeding resources, and biosafety regulation.

In short, the basic message of this book is about a comprehensive theory and proposal of intellectual property, biosafety, and business regulation covering any kind of germplasm.

The positive externality expected as a result of the application of the system is the recovery of productivity rates in crops and food biodiversity through the massive use of genetic resources and their application to the development of new living matter for feeding, while protecting the interests and rights of all players without exceptions at the same time.

In other words, this is a new paradigm based on the promotion of innovation for "Agriculture 4.0."

Buenos Aires, Argentina   Miguel Angel Rapela

Acknowledgements

This book is the result of dozens of years of experience at the domestic and international levels, both in the private and the public sectors, in connection with plant innovation.

I have a huge debt of gratitude toward hundreds of colleagues and experts from universities, research centers, and seed companies all over the world. Aware that I am not mentioning countless institutions, I specifically wish to thank my colleagues at the Intellectual Property Committee and the Working Group on Plant Breeding and Innovation of the International Seed Federation for the opportunity to discuss many of the issues included in this work.

I would like to acknowledge Austral University and, particularly, its Intellectual Property Centre. With this contribution, the Centre is resuming the road as an institution generating ideas and projects with domestic and international significance, and I am personally proud of this.

I would also like to thank Dr. Andrés Sánchez Herrero and Dr. Gustavo Schotz, both professors at Austral University, Argentina, for their patience in reading the original text, their remarks and suggestions, in addition to the publishing assistance they offered.

A special word of gratitude is due to two professors at Universidad del Salvador, Argentina, Dr. Juan Miguel Massot, for his assistance and valuable comments, and Dr. Jorge Viñas, for his inputs about game theory.

Contents

List of Tables

Abbreviations

AIR	Am I regulated? Voluntary system applied in the USA for the regulation of biotechnological products
BGs	Bonn Guidelines
BT	Budapest Treaty on the International Recognition of the Deposit of Microorganisms for the Purposes of Patent Procedure
CBD	Convention on Biological Diversity
CGIAR	Consultative Group on International Agricultural Research
CPB	Cartagena Protocol on Biosafety
CRISPR-Cas9	Clustered regularly interspaced short palindromic repeats. Cas-9 are natural endonuclease enzymes which are always associated with CRISPR loci
DNA	Deoxyribonucleic acid
DSB	DNA double-strand break
DSI	Digital sequence information
EDV	Essentially derived variety
EPSPS	5'-Enolpyruvylshikimate-3-phosphate synthase
GATT	General Agreement on Tariffs and Trade
GMO	Genetically modified organism
HDR	Homology-directed repair
HPVs	Heterogeneous plant varieties
INDEL	DNA base insertions or deletions
IPRs	Intellectual property rights
ITPGRFA	International Treaty on Plant Genetic Resources for Food and Agriculture
MAS	Marker-assisted selection
MATs	Mutually agreed terms
NBTs	New breeding techniques
NHEJ	Non-homologous end joining
NP	Nagoya Protocol
ODM	Oligonucleotide-directed mutagenesis
OECD	Organisation for Economic Co-operation and Development

OP	Open procedure. The procedure to be applied to facilitate access to germplasm and fairly and equitably share the benefits derived from the use of such resources, on the basis of complementarity and mutual strengthening
PBRs	Plant breeder rights of the UPOV Acts
PCPIP	Paris Convention for the Protection of Industrial Property
PCT	Patent Cooperation Treaty
PGRs	Plant genetic resources
PIC	Prior Informed Consent
PLT	Patent Law Treaty
TALEN	Transcription activator-like effector nucleases. These are artificially created restriction endonuclease enzymes
TRIPS	Agreement on Trade-Related Aspects of Intellectual Property Rights
UPOV	International Union for the Protection of New Varieties of Plants
WTO	World Trade Organization
ZFN	Zinc finger nucleases with zinc fingers. These are artificially created restriction endonuclease enzymes

About the Author

Miguel Angel Rapela is an agricultural engineer specializing in genetics and plant breeding and holds a Ph.D. in Forestry and Agricultural Sciences from the National University of La Plata, Argentina, with a focus on intellectual property and regulatory matters. He is Executive Director of Research and Consulting at the Austral University Intellectual Property Centre, and a Professor and a member of the Academic Committee for the MSc course in Intellectual Property at the same University. He is also Head of Technology Transfer for the Genomic Platform, UBATEC SA and member of the Technical Committee of the Argentine Seed Commission (CONASE).

He has been President of the Faculty of Agronomy Foundation (National University of Buenos Aires), a member of the Argentine Advisory Commission on Agricultural Biotechnology (CONABIA), Executive Director of the Argentine Seed Association (ASA) and of the Argentine Association of Plant Breeders (ArPOV), and Chairman of the Intellectual Property Committees of the International Seed Federation (ISF) and of the Seed Association of the Americas (SAA). He has published more than 120 technical papers and five books on intellectual property and regulatory matters, and has developed several new plant varieties registered at the Argentine National Seed Institute (INASE).

Chapter 1
Post-Malthusian Dilemmas in Agriculture 4.0

Abstract This chapter contains an exhaustive description of the changes that occurred in the steps of agriculture 1.0, 2.0, 3.0 to 4.0, and the general claim of the work, which is, that "Agriculture 4.0" requires adhering to a different regulatory paradigm. An analysis is presented showing that the current network of separate sums of regulatory treaties is not the adequate tool that can answer to the demand of this new scenario.

Keywords Agriculture 4.0 · International treaties · TRIPs · Convention on biological diversity · Traditional knowledge · Productive increment of crops · Innovation Rate · Malthus · Boserup

1.1 The Claim

Years ago, a graduate student from a Latin American university presented a dissertation containing a bill for the application of one of the international treaties on the access, use, and equitable benefit sharing of plant genetic resources. One of the jurors assessing the dissertation, an expert in international treaties and laws, was in ecstasy: "This dissertation should be at the desk of every lawmaker," he said to his peers. Another juror, with a biological background in this case, disagreed: "The dissertation shows that the student worked very hard, but putting the idea in practice entails the need to overcome seventy-four administrative steps to access a genetic resource." For this expert, this made it virtually impossible to access, and then use the resources. In the end, he alleged, the result would be contrary to what had been proposed.

The way in which the problem of plant genetic resources (PGRs), intellectual property rights (IPRs), and plant biosafety regulatory frameworks has been addressed, has many times gone through the extreme opposite opinions expressed above. To a large extent, regulators, administrative officers, environmentalists, and experts in all fields, not excluding philosophers, political scientists, and thinkers in general, have worked toward creating a set of conventions, treaties, and other instruments that have rendered this field nothing more than a set of documents partially or totally unrelated to each other. Also, companies specializing in the development of plant varieties or biotechnology have not fostered the adoption of clearer and simpler regulatory frameworks. Extremely complex international, regional, and domestic laws have resulted in benefitting their interests. But, in a way, the situation is kind of a paradox, as only big international firms have the capabilities required to overcome the obstacles of regulatory and intellectual property frameworks. In turn, the developments of small companies or official institutions have been left out. In addition, almost nobody has recognized the wealth of traditional knowledge and previous work dating back thousands of years of empirical selection made by farmers all over the world. Finally, the scarce involvement in decision-making by experts-who are in daily contact with the practical reality of breeding, genetics, plant biotechnology, and plant genetic resources-, is no excuse for the responsibility in these matters. Not doing anything is also doing something.

There are a significant number of experts who can explain in great detail the scope and characteristics of each of the following instruments: Convention on Biological Diversity (CBD), International Treaty on Plant Genetic Resources for Food and Agriculture (ITPGRFA), Bonn Guidelines (BGs), Nagoya Protocol (NP), Cartagena Protocol on Biosafety (CPB), Convention of the International Union for the Protection of New Varieties of Plants (UPOV), Agreement on Trade-Related Aspects of Intellectual Property Rights (TRIPS), Patent Cooperation Treaty (PCT), Paris Convention for the Protection of Industrial Property (PCPIP), Budapest Treaty on the International Recognition of the Deposit of Microorganisms for the Purposes of Patent Procedure (BT), Patent Law Treaty (PLT), in addition to all regional and domestic laws and regulations addressing these matters.

But things are different when this huge and complex set of laws is applied to modern plant varieties (novel, distinct, uniform, and stable) and microorganisms, which may also contain biotechnological inventions and have used plant breeding resources and even traditional knowledge. It is extremely difficult to understand this situation in a global manner, considering overlapping, gaps, ambiguities, contradictions, and inconsistencies, and put it in the context of two scientific specialties—genetics and biotechnology—and a technical specialty—plant breeding—whose development and progress in the twentieth and twenty-first centuries so far has been unstoppable and exponential, while at the same time it coexists with ancestral practices in connection with the use and conservation of genetic resources by local communities (Rapela 2015).

There is no doubt that the existing bipolarity traces its origin back to a particular "big bang," as it included virtually simultaneous events that took place in two areas with no apparent connection between them.

The first was the Uruguay Round within the framework of the General Agreement on Tariffs and Trade (GATT), spanning from 1986 to 1994, which gave rise to the World Trade Organization (WTO). This Round was the most important negotiation of any kind in the history of mankind, with 125 participating countries. During the Round, the WTO was created to replace the GATT and all relevant financial matters were discussed, including the TRIPS. There was virtually no matter connected with

IPRs left outside the TRIPS, and the scope of this instrument was extended to patents and the protection of plant varieties.

Since 1994, the TRIPS marked a clear dividing line in the trend of IPR policies all over the world, and especially in the Americas. As from that date there has been a generalized application of IPRs on plant varieties and plant biotechnology with differences in scope and effects in each country (Rapela 2007; Campi and Dueñas 2015).

Another event was connected with PGRs, that is, the part of genetic diversity with current or potential value (WIPO, 2018b).[1] These are found in the wild, at times protected by local communities, at times collected and catalogued, and they are found in germplasm banks, in situ or ex situ.

Throughout the 1980s, both "common heritage of mankind" and "free access" principles in connection with genetic resources in general and PGRs in particular started to be questioned. In this context appeared the CBD, signed by more than 150 countries during the United Nations Conference on Environment and Development—which came to be known as the Earth Summit—and which became effective on December 29, 1993. The CBD document abandoned both principles and changed them for "national sovereignty" over PGRs and "fair and equitable benefit sharing" (Rapela 2000).

It was clear back then that 1994 marked a new stage, which was later confirmed by the ITPGRFA and which became effective on June 29, 2004. The main purpose of the Treaty was the preservation and the sustainable use of PGRs for food and agriculture, and it protected those two CBD principles. The BGs, completed in April 2002; the CPB; and the NP were added to this issue resulting from the CBD.

Just as beginning in 1994 the TRIPS draw that dividing line in connection with IPRs, the CBD had the same effect, in the trend of PGRs all over the world, and particularly in the Americas. From that year onward a widespread ratification of treaties was observed (Rapela 2015).

The TRIPS, essentially a product of economists, and the CBD, essentially a product of environmentalists, with all their related treaties, were based on important and clear premises and objectives. The problem is not only that the impact assessment of these regulatory instruments questions their effectiveness in general terms; much more importantly, this mixture of a significantly high number of international treaties, with their overlapping, interactions, and contradictions, may be affecting innovation in the entire agricultural industry.

For example, for some years now companies and research institutes have been adding molecular marker-assisted selection and transgenesis to standard breeding, integrating them in a series of novel plant-breeding techniques to get very specific results in less time. Many of these techniques—collectively and improperly known

---

[1] According to the World Intellectual Property Organization, the term "genetic resources" means any "genetic material of actual or potential value. *Genetic material is any material of plant, animal, microbial or other origin containing functional units of heredity.* Examples include material of plant, animal, or microbial origin, such as medicinal plants, agricultural crops and animal breeds."

as "New Breeding Techniques" (NBTs)[2]—allow for gene editing[3] by the practical use of homologous and non-homologous recombination, with the potential to create plants with agronomically valuable characters without the addition of exogenous DNA. This would be an alternative to transgenesis, which is exactly the opposite. Other technologies of this kind, use gene-engineering methods to insert new genes in plants, with the difference that they use the natural genes of the very species that have been modified (allele replacement, cisgenesis, and intragenesis). Other methods are applied to edit or cause mutations in specific locations of a plant's genome with the help of DNA artificially produced fragments or special enzymes (Meganucleases, ZFNs [Zinc Fingers], TALENS, and CRISPR-Cas9). New methods range from conventional plant grafts in GM rootstock and DNA methylation for gene silencing to techniques allowing for the full development of synthetic genomes (Rapela 2012; Rapela 2014a, b; Rapela and Levitus 2014).

The first products developed with these techniques are already commercially available. This entails legal and regulatory challenges, as it is not clear whether any plants produced with these techniques must be considered as a genetically modified organism (GMO) (Rapela 2012, 2014a, b, 2018a, b, c).

But there are further questions: how are these products protected?

In almost every country, patent law does not protect any matter that is preexisting in nature. This law-of-nature principle, which is supported—but not always observed—by the very origin of intellectual property (i.e., protection is due to products and procedures resulting from human invention and not before any such invention), means that a gene that is present in nature cannot be patented. But would it be possible to obtain a patent over the intentional change in just one nucleotide? Or two? What about three? How many nucleotides in a DNA segment must be changed for it to amount to an invention, so that such natural gene is no longer "natural"? And if the change could have been caused by the natural event of mutation, would that be patentable anyway? What if the change is epigenetic and there is no alteration of genomic information?

Some have said that patent law should not apply to living matter. In that case, we would have to protect new inventions with breeder rights stemming from the UPOV Convention, which entails—given the definition of "plant variety"—that there must be a difference in the expression of a genetic character. So, how would a novel plant variety be protected, which has been obtained via gene editing, whose phenotypical effect is the same as that of a natural mutation or that of a prior variety?

While this is happening in the scientific world, in the context of this technological progress, the current scenario open now for the access to and conservation and access of PGRs, together with their fair and equitable benefit sharing and the pertaining regulatory aspects, could be considered at least alarming. First, it is increasingly complex to access PGRs if those PGRs have not been obtained beforehand,

---

[2] "Improperly" here means that the term NBT is applied to techniques that not only are not related to each other, but some of which cannot be considered "new" at all, such as rootstocks.

[3] Gene editing: Genome editing means the process of making accurate and specific sequence changes in the DNA of cells and living organisms.

and the current imbalance has precisely affected the fair access and use in a significant manner right when biotechnology and genetic engineering have emerged.

Restrictions on the access to and use of PGRs may exert an unfavorable pressure for their preservation, as it is not possible to discard that biotechnology—especially with NBTs—may by itself generate a significant gene variability as required by plant-breeding plans for any species.

Evidence on that is eloquent: some of the transgenes for commercial use are synthetic. If and when that moment comes—which is not unthought-of—new decisive questions will emerge: What will be the usefulness and value of PGRs? Who will be interested in their preservation? What kind of matter may be protected under an intellectual property system? When is a new product a genetically modified organism? Under which conditions should that be regulated? What would be the limit to continue narrowing the genetic base of breeding? What is the important thing: physical access to PGRs, access to the information contained in PGRs, or both?

Access to PGRs is increasingly complicated, which has been acknowledged by several experts (Boyle 2003; Bragdon 2004; Ruiz Muller et al. 2010; Ruiz Muller and Caillaux Zazzali 2014; Ruiz Muller 2015; Bass 2015; Phillips 2017; Prathapan et al. 2018; Kariyawasam and Tsai 2018; Hein 2018). The magnitude of the problem may be seen in studies regarding the access to and use of PGRs in the banks of the Consultative Group on International Agricultural Research (CGIAR), showing a drop in their access and related use (Noriega et al. 2013; SINGER 2016). Local communities, which are often the only ones interested in preserving and maintaining these resources, feel out of the debate and are afraid of any laws, even of laws aimed at preserving these resources (Ouma 2017).

Along with this international legislative effort, especially during the last decade, there have been clear alarm signs about the productive increment of crops (Pingali 2006; FAO 2009; Godfray et al. 2010; Tilman et al. 2011; Foley et al. 2011; OECD/FAO 2012; Ray et al. 2013; United Nations 2016). Several studies show that feeding and providing energy to the world would require doubling agricultural production between 2010 and 2050. Attaining this goal demands a yearly 2.4% growth rate in the main crops (Ray et al. 2013). However, the Organization for Economic Cooperation and Development (OECD) reports state that the annual average growth of agriculture in 2003–2012 was 2.1%. Much worse, the OECD estimates a 1.5% growth for 2013–2022 (OECD/FAO 2012). Private research is even more pessimistic as it shows that average improvements in the yield of corn, rice, wheat, and soy have a 1.2% growth, which would be half of what is needed (Ray et al. 2013).

Opportunities to increase production are many, and raising crop yield growth rates, together with their diversity and within a framework of sustainability, seems to be a rational measure. But the question that crops up is critical: if 200 years have passed and things have been done well, what is being done incorrectly now?

Innovation is in danger and, from a historical perspective, it is a fact that food biodiversity has shrunk to very risky limits. Of the 10,000 plant species that may be useful for human food, only 150 are cultivated, and three cereals (rice, wheat, and corn) account for approximately 60% of food needs in the world (Prescott-Allen and Prescott-Allen 1990; Cassman et al. 2003; Bhatti 2016).

A second fact based on the examination of a significant amount of information surveyed throughout history and worldwide is that between 32% and 39% of the differences in the yield of the main crops results from climate variations. In certain regions of the planet this figure may exceed 60% (Ray et al. 2015; Fitzgerald 2016; Benton 2017). As for wheat, it has been determined that a 1 °C increase in global temperature would cause a 4.1–6.4% yield decrease (Liu et al. 2016).

Climate change with telling effects on a scarce food biodiversity is the perfect mix for global disaster. Reversing this situation is possible, but that requires having the necessary genetic resources and using them by applying science and technology with a sustainable pattern.

> The claim in this study is that these facts are interrelated. The complex set of regulations about the development of modern plant varieties and beneficial microorganisms is the *cause* affecting the access to and use of genetic resources, as well as research and development in genetics and plant breeding. And the *effect* is the standstill in productivity growth, that is, the innovation rate.

If this situation is not reversed, it will not be possible to increase the crop yield growth rate, or crop diversity, let alone sustainable production in an environment affected by climate change.

1.2   The Malthus Versus Boserup Debate

This claim may be framed within the Malthus vs. Boserup debate.

A compulsive observer of the explosive population growth during the industrial revolution, Robert Malthus wrote the *Essay on the Principle of Population* in 1798. He had so little faith in himself that he published the book anonymously. Even if it was a propagandist work without any documentary support for its claims, the book was immediately successful and Malthus would publish five additional editions during his life, now with his name on them. He famously suggested that while population tends to grow in geometrical progression, food production only increased in arithmetic progression. So, there would be a day when population would be higher than means for survival, unless his proposal of preventive and repressive measures be applied (Malthus 1798).

The incorrectly termed "Malthus theory" has never been verified so far. Malthus was mistaken in his predictions, and criticism abounded. His most distinguished opponent was Danish economist Esther Boserup, who in 1965 and in the middle of the green revolution wrote a book entitled *The Conditions of Agricultural Growth*. With broad field experience, Boserup defended the positive aspects of a population in constant growth, claiming that the higher the number of people, the higher the workforce available. She predicted that when a critical point is exceeded, new agricultural forms evolve, which entails that as population increases, more pressure is exerted on the agricultural system, which stimulates innovation. Technological changes in the agricultural system, she alleged, allow for more productive crops and yield increase (Boserup 1965).

Boserup has been more accurate than Malthus, at least so far. Malthus, a bit because of negligence, but especially because he could not forecast what was happening in the fields of agricultural research, did not realize that the excessively extended era of species domestication based on empiricism and phenotype was reaching its end. In 1866, and after the work of Gregor Mendel, which would be recognized in the beginning of the twentieth century, started the era of plant breeding, albeit also based on phenotype. After more than 100 years, in 1970, the pheno-genotypical era of plant breeding opened up for researchers, and at a stunning speed, the twenty-first century finds us in the genomic era, in which the fine structure of the hereditary material is completely accessible.

What is remarkable about the Malthus vs. Boserup debate is that new agriculture forms have no answer in connection with productive increase, which leads us directly to a scenario of food crisis in the medium and long term.

Was Malthus ultimately right? Or is everything being done for Malthus to be right? An affirmative answer to the second question is most likely.

Many evident facts can be identified to support that claim: (a) access to and use of PGRs has been limited with multiple treaties; (b) PGRs continue to be underused; (c) phytogenetical progress has been impacted due to the incapacity to apply incremental breeding by restricting or removing the freedom to operate with various intellectual property laws and particularly with patent laws; (d) genetic breeding is mainly focused on high-consumption species and there are no incentives for research and development works in the long term on other species; (e) no pragmatic solution has been applied to the issue of the simultaneous coexistence of multiple rights in plant varieties and the access to PGRs; (f) the definition of what is and what is not patentable under the laws of the different countries is inconsistent.

Are we doing what should be done? It does not seem to be the case.

As explained, the impact of the TRIPS on IPRs and of the CBD on PGRs has been significant, and the evasive, and often confusing and inconsistent, set of all annex treaties combined, regional and domestic legislations, is acting like a spider web that seems to prevent innovating scientific and technical forces to deploy as in the past.

In the current context, there is no easy answer to how to ensure a continuing flow of science and technology—whether in the public or private sectors or in both—that allows for attaining production goals and in a sustainable manner. Nobody is against the idea that innovation is necessary, but the role played by regulatory and intellectual property frameworks is not clear in this context (Kock 2013). Nowadays, 10.8% of the world population is undernourished, representing 795 million people. This is the lowest figure since global statistics started in 1990 (Gould 2017). Considering the thorough research performed by Ray et al. (2013), the production deficit projected as a result of the decrease in the joint productivity rate for corn, rice, wheat, and soybean by 2050 is estimated at 1.136 billion tons a year. If we assumed these data as valid, by that year 30% of the world population would be undernourished, representing three billion people. This would be the highest figure since global statistics started in 1990.

> It is evident that, the way things are now, the world is facing an imminent and growing agricultural crisis in a perfect Malthusian model, in which Boserup's speculation that past a critical point new agriculture forms evolve and innovation is stimulated, might be wrong.

1.3   Agriculture 4.0

It is surprising that almost nobody realizes this, which is not something new in social behavior. Analysts such as Easterbrook (2004) and Pinker (2011) afterwards, among others, have pointed to the fact that while we are living the least cruel and violent era in the history of mankind, in which a proportionally higher number of people have access to wealth, health, food, and education, the general sense is pessimism. Here the phenomenon may be the opposite. While a considerable part of the world's population believes that the provision of food and water is infinite, hard data show that this could be false, and those resources could come to an end in the medium term.

According to Trigo (2016), it is a paradox as we are in a situation offering optimal opportunities in terms of technology and penetration of agriculture in the economy.

From a historical and technological point of view, this author alleges, up to the 1950 decade, "Agriculture 1.0" was characterized for being an intensive task, with relatively low energy intensity and low productivity. In the 1960s, "Agriculture 2.0" emerged with first-generation genetics and hybridization, the use of fossil energy (fertilizers, agrochemicals, machinery), and significant productivity increases as its main features. Then came "Agriculture 3.0," essentially defined around the intensity of knowledge in all areas, and particularly in connection with finding balance between high productivity and better environmental performance. Agriculture 3.0, stretching from the end of the twentieth century and the beginning of the twenty-first century, took advantage of the growing convergence between digitalization, the new biology, and various engineering disciplines, as the basis for the development of productive strategies that effectively combined the demands for higher productivity, efficiency, and sustainability, made by present-day society. Agriculture 3.0 consolidated bioeconomy as a vision for sustainable development, proposing a set of opportunities around new biology-based value chains, which have significantly redefined the supply/product relations between agriculture and industry (Trigo 2016). In a study exclusively focused on Argentina and making a lot of analogies, Campi (2013) maintains that agricultural development in this country has been a succession of stages in which technological models prevailed with their pertaining technologies, ways of organization, production, and interrelation among multiple players, which is perfectly consistent with the three agricultures defined by Trigo.

We can see this from a different perspective.

Contemporary economic realities result from the process of change in which the human being solves the economic problem of scarcity faced with needs and wishes. This process of change, which is emphasized and becomes evident in the most developed countries and replicated in many others, has a technical and productive matrix (Massot 2016). According to this author, understanding where the world is going as regards technical and productive matters helps to define the economic scenario to take strategic decisions in any country.

To be schematic, Massot's ideas may be connected to Trigo's, and the three most recent stages in the technical and productive evolution may be summarized in the Table 1.1. The left column shows the elements describing the productive system of Agriculture 1.0 and 2.0, and the right column shows the elements for Agriculture 3.0.

This comparison may be reproduced to identify the way in which production is carried out, as showed in Table 1.2.

Some changes may be noticed only in the transition from Agriculture 2.0 to Agriculture 3.0. Srinivas (2010) believes that during the green revolution—central paradigm of Agriculture 2.0—the academia and international agricultural research institutions played the most important role in history, and most of the research was generated in the public sector. This research was highly focused on underdeveloped or developing countries and regions. In contrast, in the biotechnological revolution—paradigm of Agriculture 3.0—the private sector took the lead over the academia, and most, if not all, basic developments were protected by rigid intellectual property structures. Research was notably focused on farming exporting countries with a significantly technified agricultural sector.

This explanation of technological and productive trends is an example of the significant change in the subject through the last five decades. Such change, will have an impact on future scenarios in very specific issues in connection with the agricultural and industrial sector all over the world, for instance, production pattern (what is produced), technical and productive system (how things are produced), and which will be the possibilities for the economic development of each country, as well as potential conflicts and matters on the agenda of international organizations (Massot 2016).

Table 1.1 Core components of Agriculture 1.0 and 2.0 versus Agriculture 3.0

Agriculture 1.0 and 2.0	Agriculture 3.0
Human resources	Human capital
Machinery	Information
Physical work	Intellectual work
Physical force	Emotional intelligence
Real and factual	Symbolic
What we produce	How we produce
Things	Concepts

Table 1.2 Productive related components of Agriculture 1.0 and 2.0 versus Agriculture 3.0

Agriculture 1.0 and 2.0	Agriculture 3.0
Mechanical work	Creative work
Established and regulated	Experimenting
Rigid and hierarchical organizations	Flexible and horizontal organizations
Autarchy	Interrelated and open
Top-down decisions	Top-down, bottom-up decisions
Reactive decisions	Proactive decisions
Linear dynamics	Nonlinear dynamics
Fear of the unknown	Accepting the challenge
Controlled risk	Basic uncertainty

In connection with seeds and biotechnology, there is another series of distinguishing elements of Agriculture 3.0, such as the existence of a very limited number of multinational companies with strong economic power, the command of the best international techniques in genetic engineering and the control—via patents—of important genes to introduce into specific varieties (Bisang et al. 2008).

However, Agriculture 3.0, which is clearly multinational in nature, overregulated in many cases, and often overprotected, seems to be part of the past. By way of example, the intellectual property of new developments and biotechnological frameworks regulating biosafety, covered different aspects and were independent, but they had a tacit interaction. As all biotechnological plant products were high regulated in terms of biosafety, their protection by intellectual property—specifically, via patenting biotechnological inventions—should be consequently high. With the arrival of NBTs and particularly gene editing, which allows for a universe of possibilities with higher accuracy than transgenesis, such interaction disappears to a large extent. The prevailing ideas in connection with biosafety regulation have to do with low to null regulation. What will the level of protection of those products be under intellectual property? Even further, will it be possible to protect them under the existing systems? And if products resulting from such scientific progress could not be protected, what would the incentive be to develop them?

The genetic and biotechnological progress of Agriculture 3.0 was mainly led by private companies. By devoting significant human and financial resources,[4] these companies applied transgenesis in practice, which modified agriculture in many countries all over the Americas. On the contrary, new genetics, especially gene-editing techniques, is mainly led by public academic centers, and now the ones who lead the path for the future marketing of new products are researchers.

Agriculture 3.0 seems to have passed by at full speed, and we are now in Agriculture 4.0 with different key elements, as shown in Table 1.3.[5]

---

[4] A leading international seed company even had 15 Nobel prizes among its staff. The experimental development and regulations to obtain the commercial authorization for a single transgenic event in plants requires an investment of around USD 100 million.

[5] The concept of "Industry 4.0" or also "fourth Industrial Revolution" is widespread. These terms

Table 1.3 Core components of Agriculture 3.0 versus Agriculture 4.0

Agriculture 3.0	Agriculture 4.0
Private	Public
Companies	Universities
Monopolistic knowledge	Democratization of knowledge
Invasive genetic changes	Accurate genetic changes
Stochastic genetic changes	Site-addressed genetic changes
Physical access to the resource	Access to information
Regulated	Unregulated
Exclusive licenses of patents over techniques	Non-exclusive, accessible licenses, if the technique is patented
Patents over products	In many cases, outside the patent protection scope
Very difficult technically	Very easy technically
Very expensive	Affordable to cheap
Slow	Quick
Transgenic	Nontransgenic
Detectable	Undetectable
Applicable to high-margin crops	For all crops

The innovation implied by the new technologies of Agriculture 4.0 is self-evident. While this book is about the environment of new plant breeding technologies and their relationship with PGRs and biosafety, the concept of "Agriculture 4.0" is recognized as highly dynamic, in rapid evolution, and applied to many areas connected with what is known as "intelligent agriculture." For example, the concept is associated with data automation, data mining, fully-integrated production processes, and intelligent digital ecosystems for agriculture (ICT 2015; Adam 2016). That is why the term "innovation" is intended to mean phenomena that are important in themselves, but which are often elusive (NAP 2017).

According to the Dictionary of the Spanish Royal Academy, "paradigm" is a theory or set of theories whose core is accepted without questioning and which provides the basis and model to find solutions and make progress in knowledge. Kuhn (1962) made a philosophical reflection on this aspect and proposed—not without deep criticism—that science has periods of "normal science" characterized by the existence of a paradigm. During periods of normal scientific research, scientists are trained to accept and share only one paradigm, a compact scientific construction that is a "standard" or "pattern." Within the framework of the paradigm, scientists extend the application range of their techniques, solve problems existing in their field, and become "conservative," to the extent that they are not recognized for their originality, but for their adherence to the dominant "paradigm." The tenacity of "normal science" periods is evidenced by the resistance to any kind of change

---

reflect a new way of organizing and improving the efficiency of means of production via interconnected machine and systems where information is the link for processes. The term Agriculture 4.0 as used in this text only means what is defined here.

of the dominant paradigm, and achievements accumulate and are integrated to train new ranks of scientists. The paradigm, said Kuhn in subsequent revisions of his work, is a novel way of solving a scientific issue, and it is an achievement that could start with a hypothesis, then theory, and maybe law.

How do scientific ideas change? Or how is the shift from one paradigm to another carried out?

While positivists and inductivists—mainly connected with the rationalist school of Karl R. Popper—believe that new scientific ideas result from logical arguments, verification, or falsification, Kuhn thinks that periods of "normal science" that form the base of and build a paradigm, at a certain point become unable to solve anomalies—new, unforeseen, unexpected events—which crop up. Then, those periods are interrupted by abrupt changes and are replaced by theories amounting to "scientific revolutions" in which a new paradigm is reelaborated, replaced, and rebuilt. Because, as Kuhn stated, the scientific community is basically conservative in nature, innovation—or the innovator—is a disruptive agent and there is always resistance against change, as no group—no matter their origin—accepts that the foundation on which their knowledge or activities are based be removed. A paradigm does not disappear or fall so that another paradigm emerges; it is required that another paradigm emerges and takes the place of the former one in the form of alternative, original, simple, or comprehensive interpretations.

Kuhn developed his ideas exclusively based on academic experience and applied them for the development of ideas in laboratories, universities, and other scientific centers. As already explained, Agriculture 3.0 was strongly supported by private research and what is remarkable is that those behaviors of academic life are transferable and assimilable to the private scenario.

In 2007, Nobel Peace Prize winner and key author of the "green revolution" Norman Bourlag said that the last 50 years have been the most productive period in the history of agriculture, and also that the challenge for the next 50 years is huge—population will increase 60–80%, and with it the need to generate renewable sources of energy; and to top it all, more than half of the 800 million people who suffer hunger in the world are small farmers who cultivate marginal, ecologically sensitive land in developing countries. According to Bourlag, there is no doubt that "having the science and the technology to have an impact on the protection of these fragile environments is one of the biggest challenges of the 21st century" (Borlaug 2007). It is difficult to disagree with this.

And if feeding and providing energy to nine billion people in an environmentally sustainable manner by 2050 seemed to be the highest challenge in the history of mankind, this is even more complex, as new estimates show that population growth rates in Africa and Asia could be higher than projected some years ago. The United Nations (2015) estimates 9.725 billion by 2050, in which no *plateau* would be reached, as was originally expected, but population would continue growing to reach 11.213 billion by 2100. Due to this combination of reasons—more mouths to feed, more energy to move the country, limitation or decrease of cultivable land and water available for crops—it is no surprise that food safety is at the top of the international political agenda (Ainsworth 2015; Benton 2017; Hodson 2017; Gould 2017).

To provide food and energy while preserving the environment, it will be necessary to make a deep analysis of the causes that have led to a significant decrease of genetic progress in the most important cultivated species and, of course, to find ways to reverse that trend.

The complex set of conventions, treaties, and acts regulating modern plant varieties, presented in greater detail in Chap. 2, were in some cases, drafted before Agriculture 1.0, and in other cases, mostly during Agriculture 1.0, and realized, drafted, and enacted during Agriculture 2.0 and 3.0.

It is unlikely that this could be successful, but even worse, this could be counterproductive.

> Agriculture 4.0 wants to adhere to a different regulatory paradigm in its broadest conception, and it does not seem that the current set of individual elements is the tool that could respond to this new scenario.

This is the post-Malthusian dilemma in Agriculture 4.0, which requires analysis and a proposal of a solution that will be presented and developed in the following chapters.

References

Adam U (2016) Farming 4.0′ at the farm gates. CEMA European Agriculture Machinery. Updated, July 30, 2016

Ainsworth C (2015) Agriculture: a new breed of edits. Nature 528:S15–S16. https://doi.org/10.1038/528S15a

Bass K (2015) The battle over plant genetic resources: interpreting the international treaty for plant genetic resources. Chic J Int Law 16(1):Article 7. Available at: http://chicagounbound.uchicago.edu/cjil/vol16/iss1/7

Benton TG (2017) Food security. In: Thomas B, Murray BG, Murphy D (eds) Encyclopedia of applied plant sciences, 2nd ed. volume 2, Breeding genetics and biotechnology. Elsevier, Academic Press, pp 19–22

Bhatti S (2016) Use it or lose it: the international treaty provides access to key plant breeding material. European Seed 3(4):19–22

Bisang R, Anlló G, Campi M (2008) Una revolución (no tan) silenciosa. Claves para repensar el agro en Argentina. Desarrollo Económico 48:165–207

Borlaug N (2007) Prologue. In: Krattiger A, Mahoney RT, Nelsen L, Thomson JA, Bennett AB, Satyanarayana K, Graff GD, Fernandez C, Kowalski SP (eds) Intellectual property management in health and agricultural innovation: a handbook of best practices. MIHR/PIPRA, Oxford/Davis. Available online at www.ipHandbook.org

Boserup E (1965) The conditions of agricultural growth: the economics of agrarian change under population pressure. Allen & Unwin, London. Reprinted as: Boserup, Esther (2005) The conditions of agricultural growth: the economics of agrarian change under population pressure. New Brunswick, New Jersey

Boyle J (2003) The second enclosure movement and the construction of the public domain. Law Contemp Probl 66:33–74. http://creativecommons.org/licenses/by-sa/1.0. It is also available at http://www.law.duke.edu/journals/66LCPBoyle

Bragdon S (ed) (2004) International law of relevance to plant genetic resources: a practical review for scientists and other professionals working with plant genetic resources. Issues in genetic resources, no 10, march 2004. International Plant Genetic Resources Institute, Rome

Campi M (2013) Tecnología y desarrollo agrario. In: Anllo G, Bisang R, Campi M (eds) Claves para repensar el agro en la Argentina. Buenos Aires, Eudeba, pp 97–152

Campi M, Dueñas M (2015) Intellectual property rights and international trade of agricultural products. World Dev 80:1–18

Cassman KG, Dobermann A, Walters DT, Yang H (2003) Meeting cereal demand while protecting natural resources and improving environmental quality. Annu Rev Environ Resour 28(1):315–358

Easterbrook G (2004) The progress paradox: how life gets better while people feel worse. Random House Trade Paperback Edition

FAO (2009) Global agriculture towards 2050. FAO, Rome

Fitzgerald T (2016) The impact of climate change on agricultural crops. In: Edwards D, Batley J (eds) Plant genomics and climate change, pp 1–14

Foley JA, Ramankutty N, Brauman KA, Cassidy ES, Gerber JS (2011) Solutions for a cultivated planet. Nature 478:337–342

Godfray HCJ, Beddington JR, Crute IR, Haddad L, Dl L (2010) Food security: the challenge of feeding 9 billion people. Science 327:812–818

Gould J (with infographic by Ashour M) (2017) A world of insecurity. Nature 544:6–7

Hein T (2018) No access, no benefits – part 3 – the view from academia. European Seed, posted on December 12th, 2018 by Treena Hein. International News, Regulatory. 5(4). https://european-seed.com/2018/12/no-access-no-benefits-part-3-the-view-from-academia/

Hodson R (2017) Food security. Nature 544:5

ICT (2015) Agriculture 4.0 The Internet of things boosting regional innovation. Networking Sessions Programme. Innovate, Connect, Transform, European Commission. ICT 2015, 20 to 22 October, Lisbon, Portugal

Kariyawasam K, Tsai M (2018) Access to genetic resources and benefit sharing – implications of Nagoya protocol on providers and users. J World Intellect Prop 21(5–6):289–305

Kock M (2013) Adapting IP to an evolving agricultural innovation landscape. WIPO Magazine. http://www.wipo.int/wipo_magazine/en/2013/02/article_0007.html

Kuhn TS (1962) The structure of scientific revolutions. Chicago, IL: University of Chicago Press

Liu et al (62 authors) (2016) Similar estimates of temperature impacts on global wheat yield by three independent methods. Nat Clim Chang 6:1130–1136

Malthus R (1798) An Essay on the Principle of Population, as it Affects the Future Improvement of Society with Remarks on the Speculations of Mr. Godwin, M. Condorcet, and Other Writers. London Printed for J. Johnson, in St. Paul's Church-Yard

Massot JM (2016) Biotecnología, desarrollo económico e inserción internacional. Dos propuestas para el caso argentino. Agenda Internacional 36:56–68

NAP (2017) Advancing concepts and models for measuring innovation: proceedings of a workshop. National Academies of Sciences, Engineering, and Medicine. The National Academies Press, Washington DC. https://doi.org/10.17226/23640

Noriega IL, Halewood M, Galluzzi G, Vernooy R, Bertacchini E, Gauchan D, Welch E (2013) How policies affect the use of plant genetic resources: the experience of the CGIAR. Resources 2:231–269

OECD/FAO (2012) OECD-FAO agricultural outlook 2012–2021. OECD Publishing and FAO Rome. https://doi.org/10.1787/agr_outlook-2012-en

Ouma M (2017) Traditional knowledge: the challenges facing international lawmakers. WIPO Magazine, February 2017. This article is based on the keynote address by Dr. Ouma at the WIPO Seminar on Intellectual Property and Traditional Knowledge in Geneva, Switzerland, in November 2016

Phillips PWB (2017) Ownership of plant genetic resources. In: Thomas B, Murray BG, Murphy D (eds) Encyclopedia of applied plant sciences, 2nd ed, Vol 2, Breeding genetics and biotechnology. Washington, DC. Elsevier, Academic Press, p 28

Pingali P (2006) Westernization of Asian diets and the transformation of food systems: implications for research and policy. Food Policy 32:281–298

Pinker S (2011) The better angels of our nature: why violence has declined. Viking, New York. ISBN 9780670022953

Prathapan D, Pethiyagoda R, Bawa KS, Raven PH, Rajan PD and 172 co-signatories from 35 countries (2018) When the cure kills: CBD limits biodiversity research. Science 360 (6396): 1405–1406

Prescott-Allen R, Prescott-Allen C (1990) How many plants feed the world? Conserv Biol 4(4):365–374

Rapela MA (2000) Derechos de propiedad intelectual en vegetales superiores. Editorial Ciudad Argentina, Buenos Aires, 466 pages

Rapela MA (2007) Schemes for intellectual property. Seed News Magazine 11(1):20–23

Rapela MA (2012) Post-transgénesis: nuevas técnicas de mejoramiento vegetal. AGRONEXO, Revista de la Asociación de Ingenieros Agrónomos de Uruguay 1(3):8–9

Rapela MA (2014a) La era post transgénicia y el desafío de las nuevas técnicas de mejoramiento. Actas del Seminario organizado por el Instituto de Genética "Ewald Favret" del INTA Castelar en conmemoración del 45 Aniversario de la Sociedad Argentina de Genética y los 50 años de la creación del híbrido de maíz forrajero. Castelar, 5 de diciembre 2014

Rapela MA (2014b) Post-transgenesis: new plant breeding techniques. Seed News Magazine 18:14–15

Rapela MA (2015) The adoption of conventions and treaties related to genetic resources and intellectual property issues: current situation and status in the SAA region. Conference at the 5th Congress of the Seed Association of the Americas. Cancún, México, September 10th, 2015

Rapela MA (2018a) Gene editing and CRISPR-Cas. Seed News Magazine 22:12–16

Rapela MA (2018b) Metodología de CRISPR, aspectos legales y regulatorios. Actas XI Congreso Nacional de Maíz, Mesa de Genética y Mejoramiento Genético Vegetal, pp 266–270

Rapela MA (2018c) Edición Génica mediante sistemas CRISPR/Cas. AGROPOST CPIA-Consejo Profesional de Ingeniería Agronómica 155(abril-mayo):11–13

Rapela MA, Levitus G (2014) Novas técnicas do melhoramento. In: Anuario da ABRASEM. Associação Brasileira de Sementes e Mudas, pp 29–32

Ray DK, Mueller ND, West PC, Foley JA (2013) Yield trends are insufficient to double global crop production by 2050. PLoS One 8(6):e66428. https://doi.org/10.1371/journal.pone.0066428

Ray DK, Gerber JS, MacDonald GK, West PC (2015) Climate variation explains a third of global crop yield variability. Nat Commun. https://doi.org/10.1038/ncomms6989

Ruiz Muller M (2015) Genetic resources as natural information: implications for the convention on biological diversity and Nagota protocol. Taylor & Francis Ltd, London, 170 pages

Ruiz Muller M, Caillaux Zazzali J (2014) Propiedad Intelectual y acceso a Recursos Gnéticos en un ambiente altamente politizado y "desinformado". Anuario Andino de Derechos Intelectuales 10(10):317–332

Ruiz Muller M, Henry Vogel J, Zamudio T (2010) La lógica debe prevalecer: un nuevo marco teórico y operativo para el Régimen Internacional de Acceso a RGV y Distribución Justa y Equitativa de Beneficios. Documentos de Investigación 5(13). http://www.spda.org.pe

SINGER (2016) Data base of the System-Wide Information Network for Genetic resources (SINGER) Accessed, September 2016 at http://www.singer.cgiar.org

Srinivas KR (2010) Population and demographic change. In: Spring, Ursula Oswald, Ada Aharoni, Ralph V. Summy, and Robert Charles Eliot (eds). Peace studies, public policy and global security, Volume VI, p 38

Tilman D, Balzer C, Hill J, Befort BL (2011) Global food demand and the sustainable intensification of agriculture. Proc Natl Acad Sci U S A 108:20260–20264

Trigo E (2016) Potential productivity increases in the Argentine agri-food production. Publication
    of the "Grupo de Productores del Sur", GPS
United Nations, Department of Economic and Social Affairs (2016). Population Division,
    Population Estimates and Projections Section. Accessed June 2016 at: http://esa.un.org/unpd/
    wpp/index.htm
United Nations, Department of Economic and Social Affairs, Population Division (2015) World
    Population Prospects: The 2015 Revision, Key Findings and Advance Tables. Working Paper
    No. ESA/P/WP.241

Chapter 2
The Regulatory Tangle

Abstract This chapter contains an exhaustive description and evolution of the main international treaties on patents, breeder's rights, genetic resources, and biosafety. The difficulties of adapting these frameworks to the development of new technologies, especially gene editing, are considered. Five real gene-editing cases are analyzed in the current scenario of separate treaties and regulations.

Keywords Regulatory frameworks · Intellectual property rights · Patents · Plant breeder rights · Biotechnological inventions · Plant genetic resources · Biosafety · Gene editing

2.1 Regulatory Frameworks

Plant breeding is one of the activities that humans have been engaged in since the beginning of mankind. Archeological records show evidence indicating that the first attempts to domesticize plant species started approximately 10,000 years before Christ. Since that time, humans have selected plants for consumption and support, so much so that no commercial plant species may now be considered a product of nature as such. What we eat today has been altered or modified to a larger or lesser extent. Humans have transformed everything. That transformation took place first during an extended period of species domestication based on empiricism and selection by phenotype. In 1866, and based on the work of Gregor Mendel, began the scientific era of plant breeding, but also based on phenotype, which was a distinguishing feature of Agriculture 1.0 and 2.0. It was only 100 years afterward, in 1970, when the pheno-genotypical era of plant breeding opened up for researchers in Agriculture 3.0, at a dramatic speed. With the knowledge of genes and their expression mechanisms, the twenty-first century finds us in a genomic era, in which it is possible to access the full sequence of the hereditary material and which has led us fully into Agriculture 4.0 (Rapela and Levitus 2014).

However, strictly speaking, in plant breeding one stage does not replace another; what actually happens is that breeders use techniques based on genetics and on the knowledge of characters dominated by genes and their interaction with the environment. This kind of plant breeding may be classically divided as follows (Rapela 2014a; Georges and Ray 2017):

- Conventional breeding: It is successful on simple-inheritance characters, high heritability, and low phenotype–environment interaction, but its success is limited when these conditions are not met.
- Modern breeding: Based on Marker-Assisted Selection (MAS), it is highly useful to identify the genes controlling mainly qualitative features and to introduce features from new germplasm sources.
- Transgenesis: Based on the transfer of exogenous DNA, it allows for the possibility of jumping the species barrier to search for variability.

For some years now, companies and research institutes have combined these stages and entered a new post-transgenic era, working in a series of techniques connected with modern biotechnology—generally known as "New Breeding Techniques" (NBTs)—and which are starting to be part of breeding programs, with different purposes and scopes. Controversies in modern biotechnology (particularly GMOs) have caused certain not-so-new techniques, such as grafts or cisgenesis/ intragenesis, to be also included in the list of NBTs. "NBT" is an arbitrary name, created at the request of the Joint Research Centre (JRC) of the European Commission, made to a disciplinary group that drafted one of the first comprehensive documents on the subject (Lusser et al. 2011) and is more focused on the identification of regulatory, instead of scientific, matters (Rapela and Levitus 2014; Rapela 2014a, b).

NBTs enable all of the following: (a) to generate changes in specific regions in the genome (genome editing); (b) to transfer limited DNA quantities among genotypes; (c) to modify features without doing any changes in the genomic sequence via epigenetic alterations; (d) to insert specific DNA sequences among sexually compatible individuals; (e) to silence gene expression; (f) to build genomes de novo.

NBTs are a set of multiple technologies including techniques that are different in many aspects: how DNA is introduced, what type of change is caused, and whether any such change is permanent or provisional. Given this heterogeneity, NBTs may be grouped in different ways and according to multiple criteria and here we apply the categorization made by Rapela and Levitus (2014) and Rapela (2014a, b, 2018a, b, c):

- Group 1: Directed mutagenesis techniques, whereby changes in a specific region of the plant genome are made, including particular mutations and directed gene insertion. Example: Oligonucleotide Directed Mutagenesis (ODM), Genome Editing (Meganucleases, ZFN-1, ZFN-2, TALENs, CRISPR-Cas9), Agroinfiltration (agroinfiltration proper and agroinoculation). Unlike random mutagenesis, these techniques allow for the introduction of specific mutations in a directed manner, preserving the integrity of the rest of the genome.
- Group 2: Variations of the genetic transformation technique currently used. Example: Allelic replacement, cisgenesis, and intragenesis, focused on the intro-

duction of genes of the same species or related species. They are useful for the transfer of wild relatives, the overexpression of genes for resistance against diseases, and tolerance to abiotic stress, and the silencing of endogenous genes. This group can also include grafts on a transgenic foot.

- Group 3: Techniques entailing the generation of a transgenic in intermediate stages of the development of a product (the product does not contain any transgenes). For example, reverse engineering, dependent RNA-directed DNA methylation.
- Group 4: Agroinfiltration techniques, including techniques mediated by Agrobacterium or viral vectors, and the floral-dip method. These are techniques employed in research, especially to study the impact of protein overexpression, the function of genetic elements, and the plant–pathogen interaction.
- Group 5: Engineering of biological components that do not exist in nature, or reengineering of existing biological elements, for example, synthetic genomics.

In this entire set, group 1 clearly stands out from the rest, and particularly gene-editing techniques, the logic of which is conceptually identical for the types of genomes developed so far: Meganucleases, ZFN-Zinc Fingers, TALENS, and CRISPR-Cas. What these techniques made possible first is the DNA double-strand break (DSB) by a DNA endonuclease enzyme (FokI in the case of ZFN and TALEN, and mainly Cas9 for CRISPR-Cas). Depending on each technique, the endonuclease is addressed to a specific target in the DNA strand, which would break the DNA. The double break of the strand in the specific spot is subsequently repaired by DNA natural processes via two alternative mechanisms: (a) an effective, but error-prone, method by recombining non-homologous end joinings (NHEJs), or (b) a less effective method, but highly accurate, which is homology-directed repair (HDR). The later, requires a mold (homologous) which is capable of recognizing by the DNA sequences by complementarity on both sides of the cut. Repairing the DSB via the NHEJ mechanism may lead to the introduction of little DNA base insertions or deletions (INDEL) or, if via the HDR mechanism, the full replacement of the target sequence via the mold introduced (Ceasar et al. 2016; Egelie et al. 2016; Mojica and Montoliu 2016; Samanta et al. 2016; Schiml and Puchta 2016; Shreya et al. 2017; Scheben et al. 2017; Puchta 2017; Urnov 2018).

Gene-editing techniques cannot obviously be applied to edit a gene that does not exist in the target organism. As a matter of fact, genetically modified, insect-resistant plants developed by introducing microorganism gene sequences so far can only be attained by transgenesis. However, with these techniques it would be possible to edit these genes, which have already been introduced, modifying their expression or introducing these genes in a specific and predetermined spot in the genome.

Gene-editing methods have evolved at a high speed, as well as their simplicity, speed, specificity, and cost reduction. The first report of an organism whose genome was edited via ZFN was published in 2000; after 2 years, via TALENs, and 2 years later via CRISPR-Cas9. But while the progress with the first two techniques was slow because of the high complexity of both, the development of the CRISPR-Cas9 technique sparked a scientific revolution given its comparative simplicity

(Jinek et al. 2012; Cong et al. 2013; Jiang et al. 2013; Mali et al. 2013; Urnov 2018) and has been applied to every type of organism with a high degree of effectiveness (Mojica and Montoliu 2016). The spectrum on which CRISPR-Cas9 has been applied includes microorganisms, plants, fish, birds, mammals, and even human beings, vindicating the "first strike" phenomenon. In other words, this means that being the first person in science in doing something on a new organism is a way of establishing historical authority. In fact, any genome of any organism can now be edited and modified (Kozubek 2016). Significant technical difficulties, such as the possibility that transgenic fragments are unintentionally incorporated into the receiving genome or that gene-editing events occur in nontarget sites, have been minimized and the safety of the CRISPR-Cas9 technique is continuously evolving (Liang et al. 2017; Scheben et al. 2017; Puchta 2017).

The original technique was enhanced via the development of a CRISPR-Cas12 system. This system uses a family of endonucleases (originally known as Cpf1), guided by RNA found in *Prevotella* and *Francisella* bacteria, which do not require a crARN transactivator (tracrARN), unlike the traditional CRISPR-Cas system containing a Cas9 from *Streptococcus pyogenes*, and which need two RNAs: crARN and tracrARN, to recognize the DNA sequence to be modified (Zetsche et al. 2015). The Cas12 endonuclease can also cut the DNA in different spots, offering more options by selecting a new editing spot. While Cas9 cuts both DNA strands on the same spot, Cas12 leaves a larger thread than the other, creating "sticky" endings that are more useful in gene-editing works, increasing the possibilities of the technique to insert new sequences on the cut site, as compared to Cas9 (Zetsche et al. 2015), which has already been tested on mammals and also on soybean (Kim et al. 2017).

More progress was obtained by using the CRISPR system with Cas13a (previously, C2c2) and Cas13b (previously, C2c6) nucleases, from *Leptotrichia wadei* and *Prevotella sp*, respectively. This system has many advantages over Cas9 and Cas12, as Cas13 enzymes do not contain any RuvC and HNH domains, which are responsible for the DNA split, so they are not useful to edit DNA, but they can edit RNA. And RNA editing does not require homologous recombination to reconstitute the molecule edited. The editing via Cas13 does not require PAM sequencing in the locus diana either, making them more flexible than Cas9/Cas12 (Abudayyeh et al. 2017). The Cas14 nuclease has also been discovered in an Archaea genome, and it is the smallest enzyme found so far and which is only a third of the Cas9's size. This enzyme allows for gene editing in very small cells and even in viruses (Harrington et al. 2018).

Modifications of the technique have been developed via alterations of the union spot of the Cas protein to the DNA double thread, permitting the cut in a single strand (nickase), which significantly reduces the possibility of cuts outside the target spot. It has also been possible to modify both cutting spots of the Cas protein so as to generate a "dead Cas" (dead Cas9 or dCas9), the result of which is a programmable protein that may be merged with other enzymes and/or regulators, permitting permanent genome editing without breaking the DNA thread. The double modification of the Cas protein so that it does not cut the DNA makes it possible to silence genes and attain transcriptional repression. In this case, the modified Cas9 protein, directed by its guiding RNA, may be directed toward the region promoting a gene, reducing, or annulling its transcriptional activity and gene expression (Scheben et al. 2017; Puchta 2017; GHR 2018; Gao 2018; Shimatani et al. 2018). Moreover, and after the finding of a new Cas9 protein coming from *Campylobacter jejuni*, it has been possible to expand the spectrum of possibilities of the technique, without limiting it to NDA editing, extending it to NRA editing. NRA is a key molecule that acts as a messenger of the information stored in the genome of a cell, translating it into protein, but there are also a considerable number of pathogens, especially viruses, exclusively made up by NRA. Potential uses of this technique are continually growing (Dugar et al. 2018; Schindele et al. 2018). Recently, progress has been made to increase the accuracy in the technique with the purpose of avoiding editing in nontarget spots (Alkan et al. 2018).

Peter Barton Hutt, an expert in legal ethics who was an analyst of the Asilomar conference of February 1975, made the following remark about the CRSPR-Cas9 technique (quoted by Kozubek 2016, page 63): *"We are simply entering a new, unprecedented era."*[1]

2.2 Protection of Products and Regulatory Framework

Nowadays, NBTs in general and gene-editing techniques in particular have sparked interest all over the world in connection with two aspects: (a) regulatory and safety issues, and (b) intellectual property. In connection with regulatory and safety issues, reports, workshops, and reviews on the matter attempt to answer a question that seems to be simple at first sight: Should the products resulting from editing techniques be regulated as transgenics? Far from being easy, the answer entails a high number of considerations, including the way in which every regulatory framework defines "transgenic" and whether the trigger for the assessment is the obtaining process or the new product.

Regarding the protection via intellectual property of editing techniques, the topic does not deviate from the general rules governing this specialty. Plant varieties produced with any of these technologies may be covered via the protection of Plant Breeder Rights (PBRs) in consistency with the various UPOV conventions, and biotechnological processes and inventions may be protected with patents, provided

---

[1] While this is not the topic of this book, it cannot go without mention that, as if anything was missing in this already exciting scenario, at least five universities and research centers (the University of California at Berkeley, the Broad Institute—a foundation focused on research which belongs to Harvard University and the Massachusetts Institute of Technology—the Vienna University, the Rockefeller University, and the Vilnius University) are fighting over the ownership of the patents on the development of the CRISPR-Cas9 technology to edit genomes in superior organisms, in the context of a business whose potential is estimated in more than 50 billion dollars. This conflict over the ownership of patents would be the main—or only—reason why developers of the technique have not received the Nobel Prize for that technique, an event which is unprecedented in any area of science.

that the requirements for those protections are met. However, in the case of patents, there might be interpretation difficulties as it is common for many domestic legislations to provide that products, as found in nature, cannot be protected. The question that requires answer in each case is: what is the level of human intervention justifying that a new product is not a product as found in nature (Rapela 2014a, b, 2018a, b, c)?

I will delve into the context of each of these issues from the perspective of regulatory matters.

2.2.1 Patents

A patent entails an exclusive right given by the Government to an "inventor," which allows the inventor to prevent any other person from manufacturing, using, selling, or importing the inventor's invention for a limited period of time (Rapela 2000). A patent may be issued for products or processes that are new, entail some inventive activity, and have any industrial application, and that provide for a new way of doing something or that offer a new technical solution to a problem. To obtain a patent, the technical information on the invention must be disclosed to the public in a patent application. Patents are not merely abstract concepts and they have an invaluable practical role in modern daily life. In compensating for ideas, patents promote innovation and new technologies in all fields (Rapela 2013; WIPO 2017).

Patents are the oldest form of protection of intellectual property (Kwong 2014)[2] and, together with copyright, trademarks, trade secrets, designs, and geographic indications, they are part of what is known as primary property rights.

Patents were originally devised to protect inanimate inventions. However, and while as early as 1781 in France and 1783 in the United States, Louis Pasteur received the first patent issued in history for a live organism (an isolated yeast free from pathogenic organisms obtained in the process of beer industrialization enhancement), the reluctance to grant patents over these organisms was a common element in all patent systems (Myszczuk and De Meirelles 2014). This standard would undergo its first significant change in 1980, when the United States Supreme Court established a new concept in terms of patentability: "The laws of nature, physical phenomena, and abstract ideas have been held not patentable. Thus, a new mineral discovered in the earth or a new plant found in the wild is not patentable subject matter." *These discoveries are "manifestations of … nature, free to all men and reserved exclusively to none."* But, the ruling continued, "micro-organism plainly qualifies as … patentable subject matter." The decision of the Court was the key element in Diamond v Chakrabarty, which was about genetically modified bacteria—which did not exist previously in nature and was created in a laboratory—

---

[2] According to intellectual property expert Robin Jacob, it is possible to date the first precedent back to 600 BC, granted for a "new type of bread." However, the history on which the new system is based begins in Florence, Venice, and England, in the years between 1420 and 1450.

useful to clean oil spills. A few years later, in 1985, the same Court in the *Ex Parte Hibberd* case recognized the patentability of sexual reproduction plants, including plant varieties, seeds, plant parts, genes, plant-breeding processes and methods (Rapela 2000).

Mainly because of the fast progress in biomedical research, many countries, especially the United States, started to issue patents for biotechnological inventions made in a laboratory, but in some cases also for natural DNA sequences and even human genes. Ethical, legal, and financial concerns were present throughout the entire process, which resulted in a new decision by the US Supreme Court in *Association of Molecular Pathology v Myriad Genetics.* The nine justices of the Supreme Court unanimously decided that a DNA segment of natural origin is the product of nature and, as such, cannot be patented. However, the so-called cDNA or complementary DNA (i.e., artificial DNA obtained via a messenger RNA mold containing only sequences codifying proteins and not intron sequences), can be patented, as it is not a natural product (Barraclough 2013; Egelie et al. 2016).

The development of recombinant DNA technologies and the possibility of creating new biological structures, particularly genic constructions that did not exist previously in nature, led to attempts to revise the criteria. Article 27, Section 5, TRIPS (1994) addressed this in particular and established that patents should be available for "any inventions, whether products or processes, in all fields of technology …, patents shall be available and patent rights enjoyable without discrimination as to … the field of technology." Also, the TRIPS established that it would be possible to exclude from patentability any "plants and animals other than micro-organisms, and essentially biological processes for the production of plants or animals other than non-biological and microbiological processes. *However, Members shall provide for the protection of plant varieties either by patents or by an effective sui generis system or by any combination thereof."* Together with this, Article 30 set forth that it would be possible to "provide limited exceptions to the exclusive rights conferred by a patent, provided that such exceptions do not unreasonably conflict with a normal exploitation of the patent and do not unreasonably prejudice the legitimate interests of the patent owner, taking account of the legitimate interests of third parties."

Such language, far from being clear in a new and cutting-edge technological area, had the opposite effect and in the opinion of Kock (2009), "from a global perspective, the area of patents for plant inventions suffers from a lack of harmonization as no other area of technology" and he listed the following:

- Some countries (United States, Australia, and Japan), allow for the protection of plant varieties via patents or breeder rights.
- In the European Union it is possible to obtain a patent for a plant, but it is not possible to claim a right over a plant variety.
- Beyond the countries mentioned, in virtually all other countries in the world patent applications over plants or plant varieties are not accepted, but in some cases patent applications are admitted for DNA sequences developed in a laboratory and which were not preexisting in nature. In fact, it could happen that this protec-

tion of a biotechnological invention would extend to the organism containing it and, as a result, plant or plant varieties, albeit not patentable as such, may violate a patent. There are no similar strategies established for native features or characteristics. Some countries only provide protection via breeder rights for a select list of plant species and, as a result, there is no intellectual protection available. This situation is often inconsistent with the requirements under the TRIPS.

This is even more complicated in the context of the research exception. A large number of modern plant varieties are now protected by a PBP and at the same time contain patented biotechnological inventions. From an experimental point of view, what is possible to do with them?

In this aspect, the nonspecific language of Article 30, TRIPS, served as the basis to apply any solution. While in the United States and in Australia there is no experimental exception to patents and, thus, no commercial plant variety containing a patented invention may be used for plant breeding without the patent holder authorization, in other countries the situation is completely the opposite, and in others the exception only applies to experimentation with academic, non-business purposes (Rapela 2006; Kock 2009; Bergadá et al. 2016). In the United States, where, as explained above, patents can be applied to the protection of the plant variety, the breeders' exception is virtually nonexistent. Even if due to this aspect the protection system is the strongest in the world, this is not necessarily reflected in the crop genetic gain or productivity rate. For example, for soybean cultivation, the genetic gain or productivity rate in the United States in the last years has been comparatively similar and even lower than countries with protection regimes exclusively based on the UPOV Convention such as Argentina and Brazil (de Felipe et al. 2016), when in the past it was the other way around (Lange and Federizzi 2009). The same has happened with the cultivation of corn (Eyhérabide 2015). Other analyses seem to have reached the same conclusions, but due to different reasons (Campi 2016a, b).[3]

The problem of the inconsistency between the breeder exception to breeders' rights and the exception to experimenting of the patents regime was the main focus of attention of the debates in the International Seed Federation and after discussing the issue for years the Federation proposed a balanced model between the rights of the patent owner and the breeder of the variety applicable to the experimenting exception. This model was unanimously approved during the Assembly of the World Seed Congress held in Rio de Janeiro, Brazil, in 2012 (ISF 2012). While the

---

[3] These studies show that the innovation rate in soybean quantified by the number of varieties registered or launched to the market in the United States and Argentina is similar in both countries, and that the adoption rate regarding new varieties is higher in Argentina, where the IPRs regime is weak as compared to the United States. The author suggests that this is an argument against the standard economic theory that supports the relationship between IPRs and innovation. However, in quantifying innovation via the number of varieties registered (and not the genetic gain rate), the fact that other factors that are not relevant in terms of innovation may have been responsible for the results encountered has been ignored. For example, no attention was paid to the fact that in Argentina the high soybean variety launch rate (and other self-pollinating plants) is but a defense mechanism devised by breeders because of the extension of the illegal seed trade.

consensus reached was significant, the correct way of implementing this model is including the breeder exception in patent laws, and so far, only Germany, France, the Netherlands, and Switzerland have done so.

The following international instruments are closely connected with patents issued (Rapela 2013; WIPO 2017):

- The PCPIP, whose first version dates back to 1883, has been ratified by 176 countries and covered three topics: (a) national treatment, (b) right of priority, and (c) general rules. In connection with national treatment, the PCPIP establishes that, as regards the protection of industrial property, the countries of the Union shall grant all nationals of other countries the same level of protection as the one granted to their own nationals. Nationals of countries who are not parties to this convention shall also be entitled to that protection, provided that they are domiciled or have real and effective industrial or commercial establishments in the territory of one of the countries of the Union.
- The PLT of 2000 became effective on April 28, 2005 and was ratified by 39 countries, with the purpose of harmonizing and speeding up procedures connected with patent applications and national and regional patents to facilitate the work for users. The PLT establishes a maximum list of the requirements which may be demanded by the offices of the contracting parties.
- The PCT of 1970, which has been ratified by 151 countries, allows to seek protection for a patent regarding an invention in many countries at the same time by submitting the "international" patent application. That application may be filed by nationals or residents of the contracting states of the PCT. The PCT does not grant patents; it is, instead, a procedure to facilitate a prior opinion about the possibilities to grant a patent, which allows to delay the national phase of the request (which is the most difficult and expensive part). For the contracting parties, this makes it easier for them to receive a written opinion or a preliminary analysis of the request to support the granting or rejection of the patent.
- The BT of 1977 has been ratified by 80 countries. This instrument allows for the deposit of microorganisms for the purposes of the procedure in connection with patents in an "international depositary authority" regardless of the location of such authority within or outside the territory of the State where the patent was applied for. When an invention is about a living organism, such as a microorganism, or its use, it is not possible to disclose the patent in writing; this may be replaced by depositing a sample of the microorganism in a specialized institution. In practice, the term "microorganism" is interpreted broadly; it includes any biological material whose deposit is necessary for disclosure purposes, especially in inventions relative to the food and pharmaceutical industries.

This international scenario must be considered together with the regional scenario. For example, in the European Union there is the European Patent Convention, administered by the European Patent Organisation and valid in all of its member states. The Convention was adopted on October 5, 1973 and became effective on December 13, 2007. Andean Community countries (Bolivia, Colombia, Ecuador,

Peru, and Venezuela) have Decision 486, which is about the Common Regime of Industrial Property covering trademarks and patents.

What is the relationship between the scope of the PCP, the PLT, and the PCT with the lack of consistency in the scope of patentability over living matter in the different countries?

Patents have been used for years as indication of the degree of innovation in science and technology in all countries. But how could this indicator be maintained when there are innovations that cannot be patented under the current legislative framework (NAP 2017)?

2.2.2 Breeder Rights

Breeder rights are an exclusive right given by the Government to a "breeder," which allows the breeder to prevent any other person from manufacturing, using, selling, or importing the breeder's plant variety for a limited period of time (Rapela 2000). Breeder rights are granted to plant varieties that are new, distinct, uniform, and stable, in addition to having a univocal denomination. To obtain a breeder right, the technical information on the plant variety must be disclosed to the public in an application containing a comprehensive description of the variety, sufficient enough so that an expert on the subject matter may identify that such plant variety is different from any other set of preexisting varieties of the same species (Rapela 2000).

Breeder rights are, together with copyrights, databases, indigenous intellectual property, industrial designs, integrated circuits, and moral rights, a set of *sui generis* IPRs.

The first ideas about protecting plant varieties originated in the nineteenth century (Rapela 2000).[4] In the twentieth century, both in the United States and in Europe, new ideas and projects were first realized via the Plant Patent Act of 1930, which was useful for the protection of asexual-reproduction varieties. Afterward, innovative concepts were developed, which led to a unique protection regime, especially applicable to the protection of plant varieties and which differentiated it from the traditional patent system designed for inanimate inventions.

Breeder rights became an international intellectual property doctrine in 1961 with the creation of the UPOV system for the protection of plant varieties and the adoption of the International Convention for the Protection of New Varieties of Plants of 1961, which was modified in 1972, 1978, and 1991 (Rapela 2000, 2016).

The subject matter protected under the UPOV Convention is the "plant variety," and the evolution of the concept, as well as its current scope, are remarkable. In the Convention versions of 1961 and 1972, the definition of variety "applies to any culti-

---

[4]The source of the ideas that many years later would materialize in an international intellectual property law doctrine applicable to new plant varieties dates back to a papal edict by Gregory XVI, dated September 3, 1833, which was a statement of ownership over new inventions in the fields of art, technology, and agriculture. The edict never became effective, though.

var, clone, line, stock or hybrid which is capable of cultivation and which satisfies the provisions of … (distinctiveness, uniformity, and stability)." This definition of "plant variety" was a matter of extensive debate and disagreement was evidenced in the UPOV Convention of 1978. Such document does not contain a definition of variety, which is not a minor issue, as this instrument failed to identify the subject matter it protects. The Convention of 1991 introduced a new definition of "plant variety" and it is the one that almost all countries have adopted in their legislations: variety "means a plant grouping within a single botanical taxon of the lowest known rank, which grouping, irrespective of whether the conditions for the grant of a breeder's right are fully met, can be defined by the expression of the characteristics resulting from a given genotype or combination of genotypes, distinguished from any other plant grouping by the expression of at least one of the said characteristics and considered as a unit with regard to its suitability for being propagated unchanged."

Based on this, the "modern" conception of variety refers to a plant grouping defined by the expression of the genotype, that is, the phenotype. The definition of variety set a significant precedent: two varieties of plants with different genotypes but whose phenotypical expression is similar cannot be simultaneously protected via breeder rights, because for UPOV-type legislations both varieties are identical.

There are many differences setting breeder rights apart from patents, but the two most important ones relate to exceptions and the third is about the right's scope. The first one is the farmer's exception—in some circumstances, the farmer is allowed to reserve and use part of the grain of a protected variety harvested with the purpose of using it as seed for the next generation. This privilege is absent in the UPOV Convention versions of 1961, 1972, and 1978, but it appears as a possibility in the 1991 version, "within reasonable limits and subject to the safeguarding of the legitimate interests of the breeder" (Rapela 2010; Dutfield and Roberts 2017).

The second difference is the breeder's exception, whereby any person may have access to any commercial protected plant variety and use it for that person's own experimental purposes (crossing with other varieties, selection, mutagenesis, etc.), without authorization by the first breeder and may register such variety under his or her name.

How are both exceptions applied in different countries if one plant variety that is protected with PBP has a patented biotechnical invention?

The answer is not clear and evinces that the last UPOV Convention of 1991 is an obsolete instrument already. This issue regarding the coexistence of rights (as explained in discussing the patent issue) has only been discussed in depth in the works of the International Seed Federation (ISF 2012), and so far has only been replicated in four European countries, where this issue is not significant enough given the scarce to null use of GM plant varieties.

In connection with the scope of breeder rights, in the conference that resulted in the first UPOV Convention in 1961, the founding members unanimously agreed on the "independence principle" as the basis of the breeder-right system. This means that a novel and distinct variety is independent of the varieties used to create such variety during the breeding process. Back then, this was a clear departure from the patent-right system based on the "*dependence principle*." This meant that if a patent was an invention developed from another patent, any such subsequent patent depended on the first one. The drafters of the UPOV Convention believed that the independence principle was key to stimulate innovation in plant breeding. Sadly, this exception was abused in such a way that the right of original breeders was impaired. Plant varieties appeared in business which were "cosmetic improvements" of scarce or no genetic value incorporated. As a result of this, the concept of "Essentially Derived Variety" (EDV) emerged. It may be recorded by the new breeder but cannot be marketed without the consent of the variety's initial breeder (Rapela 2008).

The concept of EDV was aimed at promoting original research and protecting the breeder of an initial (and not derived) variety and is easy to explain, but its determination in practice requires technical refining, trained personnel, and arbitrators who know the subject matter. It has come as no surprise that the number of cases internationally is very low, and that some decisions on the same case have been different among countries, which implies that disputes are mostly settled out of court (Overdijk 2013; Smith et al. 2013).

PBP and patents in conjunction are, at least in theory, a synergic and supplementary intellectual property system—breeder rights protect a set of entire plants, but not their components, and patents may protect parts of a plant, but not a set of them (Kock 2013). However, the interaction between the UPOV Convention and patents issued for biotechnological inventions is nonexistent in terms of treaties (Rapela 2013). Also, beyond any wishful thinking, there is little interaction between the UPOV Convention and PGR treaties. For example, all UPOV Conventions set forth that to grant breeder rights there is no requirement other than establishing novelty, distinctness, uniformity, and stability in the new plant variety, preventing in such a way a connection with PGRs that may have been used in connection with its development (Rapela 2008; Correa et al. 2015).

This international scenario has to be considered together with the regional scenario. For example, the European Union passed a breeder-right-protection system that applies to its 28 member states. The African Regional Intellectual Property Organization did the same with its system for its 17 member states. In the Andean Community (Bolivia, Colombia, Ecuador, Peru, and Venezuela), Decision No. 345 of the Protection Regime for Plant Variety Rights is applied.

2.2.3 Plant Genetic Resources (PGRs)

For many years, genetic resources in general and PGRs in particular have been considered "common heritage of humanity" and of "free access" for any person interested in their use. However, the Convention on Biological Diversity (CBD) approved during the United Nations Conference on Environment and Development—widely known as the Rio Convention—which became effective on December 29, 1993 and has been ratified by 193 States so far, has deviated from both principles. The CBD recognized the sovereignty of States over their PGRs and, based on that, access to such resources was subject to domestic regulation (and in some cases, such as in Argentina, the Federal Government reserved this power for provinces), while introducing the concept of "fair and shared benefit" between the one who delivers and the one who receives the resource (Rapela 2010, 2014c, 2015).

The CBD became the main international treaty deeply discussing biological diversity, proposing three main objectives: (a) conservation of biological diversity; (b) sustainable use of its components, and (c) fair and equitable benefit sharing obtained by the use of PGRs.

In connection with the third objective, the Bonn Guidelines (BGs) were adopted by the Conference of the Parties to the CBD in 2002 and their purpose was to implement an optional regime for the process of access to and participation in the benefits of PGRs. Afterward, since 2002 in the World Summit on Sustainable Development in Johannesburg and during 8 years, a negotiation took place within the CBD to adopt an international regime promoting and protecting fair and equitable benefit sharing derived from the use of PGRs. Finally, on October 29, 2010, at the tenth meeting of the Conference of the Parties, which took place in Nagoya, Japan, the NP was adopted; its full title is "Nagoya Protocol on Access to Genetic Resources and the Fair and Equitable Sharing of Benefits Arising from their Utilization to the Convention on Biological Diversity" (Rojas and Lorena 2013).

In connection with the third objective of the CBD, the NP became effective on October 12, 2014 and as of January 2017 it has 94 ratifications. The NP is an agreement of the executing parties with the purpose of establishing legal certainty and transparence both for suppliers and PGRs users. The Protocol introduced two innovative points: (a) a series of specific obligations that each party must undertake to ensure users' compliance with the legislation within their jurisdictions or any national regulatory requirements of the party providing the PGRs, and (b) the parties' commitment to comply with the mutually agreed-upon cooperation conditions (Greiber et al. 2012).

In promoting the use of PGRs and the pertaining traditional knowledge, in addition to strengthening opportunities to share fairly and equitably any benefits deriving from their use, the NP was ultimately intended to generate incentives to use PGRs. And that use would contribute to the conservation and sustainable use of biological diversity. The NP directly results from Article 15 of the CBD and applied a system of access to PGRs and of fair and equitable benefit sharing on two instruments: (a) the Prior Informed Consent (PIC), and (b) the negotiation of Mutually Agreed Terms (MATs). The PIC was based on the principle that each state is sovereign in connection with its PGRs and their access, and there must be official authorization from the country where the resource is obtained. MATs refer to the contract containing the relationship between the parties. MATs cover not only the conditions of physical access to resources, but also how and when benefits will be shared, as well as any limitation or restriction that may be applicable in the use and/or transmission of such resources.

However, the CBD and the NP are not the only international instruments that have dealt with shared benefit and access. The ITPGRFA also covers those topics. This instrument became effective on June 29, 2004 and so far, has been ratified by

141 countries. Unlike the NP, which is about PGRs in general, the ITPGRFA exclusively focuses on PGRs for food and agriculture (Rapela 2010).

As the ITPGRFA represented an international instrument specializing in access and shared benefit, in the terms of Article 4(4) of the NP, the ITPGRFA prevails over these provisions in the NP. These interactions are specifically mentioned in the Introduction to the NP—it is stated that the NP recognizes both the key role of the ITPGRFA and its multilateral system for the access to PGRs in consistency with the CBD (Moore and Tymowski 2005).

It has been stated that the NP is ambiguous on many points, and this entails that its extended implementation may resolve many concerns that still remain. Experts believe that this instrument should go a long way for its results to materialize. For the time being, it is difficult to foresee how the validity of the NP and its interaction with the ITPGRFA may affect the seed business itself, but it is likely that there will be effects on the companies' and public agencies' genetic-breeding plans (Van den Hurk 2011).

Another point, which is not less important, is that the NP defines the conditions and the form in which users of PGRs or traditional knowledge associated with them (i.e., seed companies and official research centers) may have access to those resources and knowledge. In establishing the general obligations for the participation in any benefits resulting from the use of such resources and knowledge, the NP requires the parties to ensure that users under its jurisdiction observe domestic legislation on the access to and use of resources, as well as the regulatory requirements of the parties where the resources or knowledge have been obtained. As stated by Kariyawasam and Tsai (2018), the NP is an unbelievable accomplishment, but its legal ambiguities have lessened its impact. Along the same line, celebrated experts on the NP point to the fact that if PGRs cannot be accessed there can be no benefit, and that it is necessary to develop simple protocols of shared benefits and access (Hein 2018).

Since the very beginning of the deliberations for the CBD and following the ITPGRFA and the NP, tension arose in connection with the access to PGRs and their relationship with IPRs, and particularly with the regimes for the protection of plant varieties under the UPOV Convention and biotechnological invention patenting (Van den Hurk 2011; Dutfield and Roberts 2017; WIPO 2018a, b). At the beginning, this tension was focused on the disclosure of the origin of PGRs or the traditional knowledge in the applications for ownership titles of plant varieties or patents. As explained in the section about patents and breeder rights, this resulted from one of the inconsistencies between the TRIPS and multiple treaties on PGRs, as the TRIPS does not include any obligation for the applicants for an ownership title based on breeder rights or patents in connection with a biotechnological invention to establish that they have obtained the PIC in the event a genetic resource has been used (Rapela 2010; Correa et al. 2015). While the TRIPS define patentability requirements, several countries require, under the CBD and now the NP, the disclosure of the source and genetic origin of biomaterials. And in some cases, failure to disclose results in the rejection or invalidity of the patent (Kock 2009).

The potential inconsistency between the TRIPS and the CBD is increased if one considers that the genetic origin is often difficult, if not impossible, to determine. For example, under the CBD, "country of origin of genetic resources" means the country that possesses those genetic resources under in situ conditions and that "country providing genetic resources" means the country supplying genetic resources collected from in situ sources, including populations of both wild and domesticated species, or taken from ex situ sources, which may or may not have originated in that country.

The problem is that genetics is not limited to certain countries, and genetic origin transcends time and frontiers (Kock 2009). Therefore, the meaning of "country of origin," under the CBD, is impractical or actually impossible. Suggestions have been made for the term to be replaced with "source of origin," that is to say that the applicant be required to state where the material was obtained (ISF 2003; Rapela 2014c). An overwhelming proof of this mistake in the CBD is, among others, the recent finding of multiple centers of origin of cultivated corn that not only include Mexico, but also Brazil and Bolivia (Kistler et al. 2018).

The CBD and the implementation of the NP have been a source of much concern among breeders, particularly in connection with legal issues, excessive administrative load, and uncertainty in connection with the resulting obligations. In connection with this, the ITPGRFA, with its multilateral and accessible system of access to PGRs (as opposed to the bilateral and narrow system of the CBD), has been an instrument that considered this type of realities (Bhatti 2016; Prathapan et al. 2018).

When each of the PGRs treaties was prepared, the underlying idea was that PGRs were an unexplored gold mine that multinational medicine, pharmaceutical, and seed companies wished to exploit with the purpose of appropriating the nature and the work of indigenous communities, via IPRs, with the subsequent purpose of selling the products at extremely high prices for the inhabitants of the countries from which those resources came (Overmann and Scholz 2017). It comes as no surprise, then, that the facilitated access to PGRs that prevailed before the CBD diluted and was replaced by a complex administrative regime that is sometimes difficult to go through, discourages, and sometimes even prevents its use, and it ultimately debilitated the first and foremost purpose of the CBD—the conservation of biological diversity (Boyle 2003; Bragdon 2004; Ruiz Muller et al. 2010; Ruiz Muller and Caillaux Zazzali 2014; Ruiz Muller 2015; Bass 2015; Overmann and Scholz 2017; Phillips 2017; Prathapan et al. 2018; Kariyawasam and Tsai 2018; Hein 2018). What has actually happened, absolutely contrary to the purposes of the CBD, the NP, and the ITPGRFA, is a decrease in the rate of access to and use of resources, according to studies made by the Consultative Group on International Agricultural Research (CGIAR) (Noriega et al. 2013; SINGER 2016).

This issue can be analyzed with further background—the CBD, the NP, and the ITPGRFA were drafted at times when the physical access to the genetic resource was the only way in which the resource could be used. With the progress of "Agriculture 3.0" and the emergence of "Agriculture 4.0," what is important now is genetic information, and not merely the tangible object, be it a seed, a flower, or a fruit.

PGR treaties, as drafted, are out-of-date, and this became evident in the 13th Conference and Meeting of the Parties to the CBD and the NP, which took place in Cancun, Mexico, in December 2016. The organizers set aside some space to discuss "unresolved matters," including the definition of "synthetic biology" (a topic that had been discussed for the last 6 years), and the potential consideration of the use of "Digital Sequence Information" (DSI) on PGRs under the NP. What happened in that meeting was that "synthetic biology" became one of the key topics in the Mexico meeting and the other parallel events, and DSI emerged as a growing problem, but so far separate from "synthetic biology" (ICC 2017; Prathapan et al. 2018).

"Synthetic biotechnology" is the unique language adopted by the CBD to cover, among others, any products resulting from NBTs and particularly from gene-editing techniques. "Synthetic biology" was included in the agenda after the meeting. For the time being, the definition prepared by the former Group of Technical Experts was approved: synthetic biology is "a new development and a new dimension of modern biotechnology combining science, technology, and engineering to facilitate and accelerate the understanding, design, redesign, manufacture, and/or modification of genetic materials, living organisms, and biological systems" (CBD 2016).

Later, the meetings of the Subsidiary Body on Scientific, Technical, and Technological Advice (SBSTTA) of the CBD, which took place in Montreal in July 2018, discussed the topic again, and gene editing was again discussed as a potential component of synthetic biology (CBD 2018a).

While at the CBD meetings there is still no agreement as to whether or not the NP includes products of "synthetic biology," there is no doubt that NBTs and particularly gene-editing techniques such as CRISPR are redefining the ways to access and use PGRs (Prathapan et al. 2018).

DSI was also an item on the CBD agenda and for that purpose a group of experts was assembled to define any potential impact of DSI use on PGRs for the three objectives of the CBD and the NP. At the Mexico meeting, the Japan and Canada representatives alleged that the DSI is not within the scope of the NP, while several other countries considered that the DSI was "equivalent" to accessing a PGR. The idea is based on considering that scientific research is using an increasing amount of information, instead of the physical genetic material, which could lead to the weakening of the NP. It was added that the value of the genetic resource lies in the information transported by the DNA sequence. Even countries such as Brazil and Namibia threatened with the possibility that PGR providers unilaterally blocked the gene sequencing of such material if a solution was not found at the international level (ICC 2017; Prathapan et al. 2018).

This is an important matter. If in the future, the bodies of the CBD establish that the research and development using DSI in connection with the PGRs are within the scope of the NP, this will have important implications in research, which should be carefully considered by seed companies and academic centers (ICC 2017).

DSI continues to be discussed by the SBSTTA. At the Montreal meeting of July 2018, it was recognized that digital information on sequences includes information on nucleic acids and protein sequences, in addition to the information resulting from specific biological and metabolic processes of the cells in the genetic resource

(CBD 2018b). However, some considerations seem to be inconsistent. For example, while it is alleged that "access to digital sequence information held in public data-bases is not subject to requirements for prior informed consent," it is considered that "the creation of digital sequence information requires initial access to a physical genetic resource, and that, therefore, a benefit arising from the utilization of digital sequence information should be shared fairly and equitably in accordance with the third objective of the CBD, the objective of the NP and Article 5(1) of the NP and in a way that directly benefits indigenous peoples and local communities conserving biological diversity so that it serves as an incentive for conservation and sustainable use" (CBD 2018b).

Another inconsistency is that while countries that are members of the CBD recognize that some members are already implementing provisions considering that DSI is equivalent to the physical access to the genetic resource (CBD 2018b), the global seed industry maintains exactly the opposite position. In one of the position papers approved in the International Seed Federation Congress in 2018, it is stated that the "ISF is strongly opposed to creating any regulatory rules relating to the access and utilization of digital/genetic sequence information (DSI) in the context of the ongoing access and benefit sharing (ABS) negotiations. Furthermore, ISF strongly discourages national and regional governments from including DSI within the scope of their ABS implementing frameworks. Regulating the access and utilization of DSI would have far-reaching negative effects on basic and applied research and breeding that supports the conservation of biodiversity and food security. In addition, regulation of DSI is inconsistent with the spirit of the CBD, is not supported by the legal definition of a genetic resource and is unnecessary to ensuring the fair and equitable benefit sharing related to the utilization of genetic resources" (ISF 2018).

In addition to all these divergent provisions and the international framework for PGRs, regional treaties also apply. For example, one of them is Decision 391 of the Andean Community on a Common Regime for the Access to PGRs, adopted on July 2, 1996 and applicable at that time in Bolivia, Colombia, Ecuador, Peru, and Venezuela. This decision was the regional result to try to implement Article 15 of the CBD on access to PGRs, which has had scarce success in the incentive and promotion of the use of PGRs.

2.2.4   Biosafety

Since before the appearance of genetically modified crops, it was understood that it was necessary for adopting countries to have procedures in place to regulate the following: (a) environmental safety of trials in laboratories, greenhouses, and fields of GMOs; (b) food safety of direct consumption of GMOs, as well as any derivative products, and (c) the commercial authorization of modern biotechnological products. Biosafety is about the assessment, handling, and communication of risks, and regulatory aspects may range from the approval of genetically modified plants for cultivation, to its consumption, import, export, up to labeling, among other things. The biosafety procedure ultimately confirms that a product resulting from modern biotechnology launched to the market is as safe as or safer than its non-genetically modified counterpart. Biosafety systems are in permanent evolution and while the idea is generalized that it is necessary to apply procedures for the regulation of the commercial release of GMOs, countries have been adopting systems that are different among themselves (Rapela 2005).

In spite of the huge efforts made by international organizations such as the FAO, OECD, and UNEP for the global harmonization of regulatory systems, the global biosafety regulation of products in biotechnology is a mosaic and is increasingly divided (Ishii and Araki 2016). If all the existing biosafety systems in the world were to be based on "solid science," the differences among systems would be minimal, and that is clearly not the case (McHughen 2012).

Biosafety systems in the United States, Canada, Australia, and many countries in South America are science-based and, therefore, decision-making is in principle predictable and effective. While not perfect, science-based biosafety systems have established frameworks providing consistent and repeatable decisions to the parties involved in the international trade of agricultural products. Disagreement has of course cropped up, and it has been dealt with and resolved, but always bearing rationality in mind to provide grounds for the regulation of international commerce of agricultural products in science. The situation is very different in the European Union. While one can argue that European regulations are based on a real concern for minimizing risks and ensuring the health and safety of Europeans, it is evident that the EU regulatory system is about the perception of risk, "acceptable" level of risk, social and economic considerations (such as the coexistence of conventional and genetically modified crops), precautionary approaches, and political agendas. Moreover, as some European countries commercially produce transgenic crops, and others even refuse to regulate the free trade of transgenic crops, there is no consistency. The voluminous evidence accumulated about commercially approved OGM safety has made no difference—in the EU the risk of rejecting products that are completely safe for health and the environment has increased or at least has not decreased over time (Rapela 2005; McHughen 2012; Falck-Zepeda et al. 2013; Smyth and Phillips 2014; NAP 2016; Tagliabue 2016).

Regulations also differ on the philosophy, the approach, the regulatory attitude, and the practical implementation of regulation itself; countries have adopted different systems in connection with biosafety regulations. The EU countries have implemented an express biosafety strategy based on the process and not on the product. And the United States has adopted a strategy that may be considered mixed in some aspects, but it is ultimately defined by the process whereby a product was developed. Canada is the only country in the world which, from the beginning of the system, adopted an express product-based strategy. Countries that have adopted process-based biosafety strategies (the overwhelming majority) apply in practice a binary regulation regime differentiating genetically modified products from products deriving from conventional technologies. They usually lack regulation regarding products deriving from conventional technologies, at the same time a complex biosafety system is applied for genetically modified products (Rapela 2005; Marchant and Stevens 2015; NAP 2016).

One of the first aspects requiring consideration is that this global gap in regulatory decision-making has generated asymmetry, asynchronous approvals of biotechnological events among countries and is affecting the international grain trade, creating challenges to feed a growing world population (McHughen 2012; NAP 2016; CAST 2018a).

A second aspect is that regulatory times have increased and costs for a genetically modified product to be approved in the main markets have rocketed. According to calculations made for 2011, the total cost for genotyping, development of genetic construction in the laboratory, transfer to organism, development of event, and commercial authorization was approximately 136 million dollars, and the time required for the biosafety regulatory process is 48 months (McDougall 2011).

Based on these data, Smyth et al. (2014) have estimated that with a 20% annual return rate, biotechnology companies are near the initial $136 million investment as late as year number 13 after approval. A one-year delay in the approval process would cost 22.7 million dollars (assuming a 20% discount rate), and a seven-year delay would cost 98 million dollars. It came as no surprise, then, that transgenic biotechnology has been left to large multinational companies, which are the only ones with resources to overcome research and development financing challenges that are so risky, significant, and long-termed. Small and medium companies, together with official research centers, have been virtually left out (Whelan and Lema 2015, 2017; CAST 2018a).

Together with each domestic regulatory framework, member countries of the CBD may adhere to the Cartagena Protocol on Biosafety.. This is an international agreement specifically focused on the transboundary movement of genetically modified living organisms resulting from modern biotechnology that may have an adverse impact on the sustainable conservation and use of biological diversity. The Cartagena Protocol on Biosafety (CPB) was adopted on January 29, 2000, became effective on September 11, 2003, and has been ratified by 103 countries so far (Rapela 2005).

The CPB objectives are stated in its Article 1: "In accordance with the precautionary approach contained in Principle 15 of the Rio Declaration on Environment and Development, the objective of this Protocol is to contribute to ensuring an adequate level of protection in the field of the safe transfer, handling and use of living modified organisms resulting from modern biotechnology that may have adverse effects on the conservation and sustainable use of biological diversity, taking also into account risks to human health, and specifically focusing on transboundary movements."

As the CPB is a binding document derived from the CBD, the definitions of its terms came to have a significant importance and were adopted by many countries. The CPB understands that a "living modified organism" is "any living organism that possesses a novel combination of genetic material obtained through the use of modern biotechnology." "Modern biotechnology" is defined as "the application of: (a) In vitro nucleic acid techniques, including recombinant deoxyribonucleic acid (DNA) and direct injection of nucleic acid into cells or organelles, or (b) Fusion of cells beyond the taxonomic family."

The point is that these definitions in the CPB were within the rationale of "Agriculture 3.0," where transgenic products could be clearly differentiated from genetically modified products. But with the appearance of NBTs and particularly with the gene-editing techniques of "Agriculture 4.0," these definitions do not withstand any rigorous analysis. Products derived from most NBTs are similar to and cannot be distinguished from conventional products, while the techniques under which such products were developed pertain to "modern biotechnology" (Whelan and Lema 2015, 2017).

The regulatory dilemma in which NBT-derived products are trapped has been clear from the beginning (European Academies 2015; Marchant and Stevens 2015; Jones 2015, 2016; Genetic Literacy Project 2016; Schuttelaar and Partners 2016; NAP 2016; Wolt et al. 2016; Ishii and Araki 2016; Whelan and Lema 2015, 2017; European Commission 2017; Georges and Ray 2017; CAST 2018b). The two key questions are: are NBT-derived products GMOs or not? Should they be regulated or not?

The US Council for Agricultural Science and Technology summarized the points of view in connection with the regulation of products resulting from gene editing in three categories (CAST 2018b): (1) gene-editing products must be regulated as conventional genetic engineering using current systems, but these systems must be progressively enhanced to better balance regulation, safety, and innovation; (2) gene-editing products must be left out of the regulatory system and must be treated just like products deriving from conventional reproduction, and (3) the systems applied to conventional-reproduction products have many gaps and must be improved to accommodate gene-editing products with a higher degree of scrutiny.

By mid-2018, Australia was making active progress to take a position and many alternatives were already being publicly considered. New Zealand, South Africa, Japan, and Korea were actively discussing the issue, but with different degrees of progress. India had stated that, initially, products deriving from NBTs could not be defined as GMOs, but there are no official supporting documents. Russia and China have communicated that they do not plan to discuss this issue (Ishii and Araki 2016; Nature Editorials 2017).

2.2.5 Regulatory Alternatives for New Products

In practice, however, there are nine clearly differentiated cases:

- Canada: As the country has product-based regulation, its regulatory system considers that the current regulation appropriately covers new developments. A product obtained via an NBT may or may not be regulated depending on whether or not it is novel in nature relative to other known products.
- United States: the Plant Protection Act authorizes the USDA APHIS (Animal and Plant Health Inspection Service) to provide regulatory surveillance in connection with certain GMOs for their introduction—that is, import, interstate

movement, and environmental release—which may present pest risk for plants under current regulations. If a GMO is consistent with the definition of a regulated article and there are plans in connection with the import, interstate movement, or environmental release of such product, an authorization (permit or notice) is needed before proceeding. If there is uncertainty about whether an agency is or is not a regulated agency, confirmation of the regulated status may be requested from the agency under a procedure known as AIR (Am I regulated?). AIR is a voluntary system—developers are not required to apply for AIR, and it is applied to any kind of agency. For this case, the purpose of the AIR process is to provide the developer with certainty about whether or not the organism obtained via gene editing is regulated. The first meganuclease AIR dates back to 2011, the first for TALEN was in 2014, and the first for CRISPR was in 2016, accumulating almost 20 information requests in a public access registry.

Together with this, in January 2017 the APHIS published and made publicly available a recommendation to revise GMO regulations, including the treatment of NBT-derived products. This was the first proposal to make a deep modification of the biosafety system since 1987 (APHIS 2017). At the same time, the FDA announced its intention to gather comments on the new plant variants developed via gene editing (FDA 2017). APHIS' proposal considered that a GMO is an organism created via genetic engineering and that "genetic engineering" means any technique using recombinant or synthetic nucleic acids with the purpose of creating or altering a genome. The proposal also defined that "synthetic nucleic acids" are nucleic acid molecules that have been chemically or otherwise synthetized or amplified, including any acids that are chemically or otherwise modified, but which may be supplemented with natural nucleic acid molecules. With this proposal of genetic engineering definition, the APHIS ratified its position to exclude any products deriving from conventional genetic breeding (including, but not limited to, breeding assisted by molecular markers, tissue cultivation, and protoplast, cell, or embryo union), and chemical or radiation-induced mutagenesis. The novel proposal was the exclusion of products that have been obtained via genetic engineering techniques, but could have been obtained via conventional breeding or chemical or radiation-induced mutagenesis. These organisms were considered as essentially identical to conventional organisms in the proposal, regardless of their development method. The proposal identified a series of *ex ante* exclusion criteria, without doing a case-by-case assessment. The exclusion criteria were as follows: An organism will not be considered genetically modified if: (1) the genetic modification of the organism was a mere suppression of any size, or the mere substitution of a pair of bases that could also be obtained via chemical or radiation-induced mutagenesis, or (2) the genetic modification of the organism entailed the mere introduction of natural occurrence nucleic acids from relative and sexually compatible species, which could cross and result in viable offspring via conventional breeding (including, but not limited to, marker-assisted breeding, as well as protoplast and tissue cultivation, or protoplast, cell, or embryo union), or (3) the organism was a "null segregant," that is, the offspring of a GMO in which the only genetic modification was the introduction of nucleic acid in the donor organism in the genome of the receiving organism, but the nucleic acid of the donor organism did not pass onto the organism's off-spring, and so the nucleic acid of the donor organism did not alter the offspring's sequence.

However, on November 7, 2017, the proposal was withdrawn. The massive amount of comments and remarks received during public consultation indicated the need for more careful review. In spite of the extended agreement as to the need to make progress with the regulation of biotechnological products and not specific biotechnological techniques, and that products of the so-called new genome-editing techniques should be regulated only if they presented pest risk for plants or risk of harmful weed, the proposal contained inconsistent definitions on certain aspects and other additional regulations that were harshly criticized by some sectors. For the purposes of clarifying its position, on March 28, 2018, the USDA's secretary communicated the "Statement on Plant Breeding Innovation," establishing that the USDA does not regulate and has no plans to regulate any plants that could have been otherwise developed via traditional breeding techniques, provided that they are not plant pests or developed with plant pests (USDA 2018).

- Argentina: Argentina was the first nation in the world that established a regulatory framework for NBTs (SAGYP 2015; Jones 2015; Whelan and Lema 2015, 2017; Duensing et al. 2018; CAST 2018b; Schmidt 2018). This was no surprise as this country has one of the oldest biosafety systems in the world, created in 1991, which has been internationally recognized by the FAO as a global reference on the matter. When the CPB was prepared, Argentina already had a fully operative regulatory system for the assessment of biotechnology. In 2012, the Argentine Advisory Commission on Agri-Biotechnology (Comisión Nacional Asesora de Biotecnología Agropecuaria, CONABIA) discussed the regulation of NBTs and after more than 2 years of debate, the discussion was exclusively focused on that the definition of "new combination of genetic material" should be the key element to decide whether or not a product derived from NBTs is considered a GMO. The regulation defined an exclusion criterion: a genetic change will always be considered a new combination of genetic material when a stable and joint insertion of one or more genes or DNA sequences that are part of a defined genetic construction have been permanently introduced in the genome of the receiving plant. The assessment is made on a case-by-case basis at the CONABIA and the applicant may ask questions before beginning with the development of the product. The opinion defines whether the new product is within the regulatory framework or not (Jones 2015; Whelan and Lema 2015, 2017; CAST 2018b).
- Israel/Chile/Brazil/Colombia/Japan. The second country in the world that has established a clear system to assess NBT-derived products was Israel. The National Committee for Transgenic Plants (NCTP) of the Ministry of Agriculture and Rural Development of Israel officially published in March 2017 the summary of the decision taken by the Committee on August 8, 2016 (Israel 2017). The decisions established that the process to develop plants via gene editing is subject to regulation, but that the product of the process is outside the scope of

regulation to the extent that the applicant shows certain requirements ensuring that no DNA sequence alien to the species was incorporated into its genome. The exclusion criteria in the Israeli regulations are very similar to those of Argentine regulations, and the process itself is also similar.

Afterward, Chile, Brazil, and Colombia followed the same path, but with different implementation. Chile did this via a methodological approach. Brazil and Colombia passed regulations (CTNBio 2018; Duensing et al. 2018; ICA 2018; Schmidt 2018). Finally, by mid-2018, Japan adopted the same criterion (Grens 2018).

The criteria adopted in the regulatory systems of Argentina, Israel, Chile, Brazil, and Japan have points in contact with the US proposal, but the process is different regarding its implementation. In the systems of Argentina, Israel, Chile, Brazil, and Japan, the implementation of exclusion is *ex post* and on a case-by-case basis, while in the US proposal and the AIR procedure, the implementation is *ex ante*, via criteria that directly disregard certain products of the regulatory system without an evaluation of them. In addition, in the system of Argentina, Israel, Chile, and Brazil, if a developer does not make a prior consultation, he or she is compulsory regulated. The only way to be left out of the regulatory system in these countries is via consultation and awaiting the exclusion opinion. In the United States, and under the AIR, if a developer is sure that his or her product is outside the scope of the regulatory system, it is possible to make no consultation and directly proceed to marketing, provided that other parties of APHIS, EPA, or FDA are not entitled to supervision based on other reasons.

- European Union: In 2016, a series of European environmentalist groups[5] asked the Ministry of Agriculture in France for an answer to four pre-court matters in what became known as Matter C-528/16: (1) Are organisms obtained via mutagenesis genetically modified organisms under Article 2 of Directive 2001/18/CE dated March 12, 2001? (2) Are varieties obtained via mutagenesis genetically modified organisms in the terms of Article 4 of Directive 2002/53/CE dated June 13, 2002, in connection with the common catalog of species varieties of agricultural plants? (3) Are Articles 2 and 3 and Annex I B of Directive 2001/18/CE, dated March 12, 2001, on the intentional release in the environment of genetically modified organisms, to the extent that they exclude mutagenesis from the scope of application of the obligations established by the Directive, a full harmonization measure restricting member States to submit organisms obtained via mutagenesis to all or part of the obligations established by the Directive or any other obligation? Or do member states have a margin of appreciation, in transposing them, which may be applied to organisms obtained via mutagenesis? (4) Is it possible to question the validity of Articles 2 and 3 and of annexes I A and I B of Directive 2001/18/CE, dated March 12, 2001, by virtue of the precautionary principle enshrined under Article 191-2 of the Treaty on the Functioning of the European Union? (OJEU 2017).

---

[5] Confédération paysanne, Réseau Semences Paysannes, Les Amis de la Terre France, Collectif vigilance OGM et Pesticides 16, Vigilance OG2M, CSFV 49, OGM: dangers, Vigilance OGM 33, Fédération Nature et Progrès.

The minister refused to comply with the claims by plaintiffs, so trade unions and farmer associations appealed to the French State Council, and the matter reached the Court of Justice of the European Union.

In the beginning of 2018 and like the cases previously described, the advocate general of the Court of Justice of the European Union (CJEU) communicated a nonbinding legal opinion. It stated that organisms developed via novel gene-editing techniques should be exempt from the European Directive obligations on GMOs. According to the expert, it is necessary to differentiate transgenesis, defined as "a genetic engineering technique that consists in inserting one or more genes from other species into the genome of another species," from mutagenesis, which "does not entail the insertion of foreign DNA into a living organism," but "nonetheless involves an alteration of the genome of a living species." While the Directive of 2001 does not expressly mention the concept of transgenesis, he stated, it "covers various techniques which could normally be described as such." This goes against plaintiffs' allegations as he states that organisms obtained via mutagenesis and those edited via CRISPR-Cas may fit those exceptions on the condition that "they do not involve the use of recombinant nucleic acid molecules or genetically-modified organisms other than those produced by one or more of the techniques/methods" by Annex I B (mutagenesis and cell fusion) (OJEU 2018a).

There were two pending decisions: in the first, the Court of Justice of the European Union had to decide on the application of the "mutagenesis exception" under Directive 2001/18/EC, which legally defines what a GMO is, and in the second one, the European Commission had to interpret whether or not the organisms obtained via NBTs are within the scope of the specific legislation on GMOs.

Contrary to what was expected and exactly against the opinion of the advocate general, on July 25, 2018, the Court of Justice of the European Union ordered that organisms genetically edited with CRISPR-Cas be regulated as if they were transgenic. In other words and for this Court, organisms obtained via genomic edition must be regulated by the same Directive for GMOs, even if they are not GMOs. According to the Court, "new techniques/methods of mutagenesis [such as CRISPR-Cas] might prove to be similar to those which result from the production and release of a GMO through transgenesis," and applying these new biotechnological tools "makes it possible to obtain the same effects as the introduction of a foreign gene into that organism," that is, the technique to obtain GMOs. The Court also held that in that case those rules would not apply, as the European Union precautionary principle could be violated (OJEU 2018b). The international scientific community reacted very critically against this decision.[6]

_____

[6] "A terrible, inconsistent, and incomprehensible ruling," and "Europe misses another innovation, scientific development train," said Lluís Montoliu, National Center of Biotechnology (CNB-CSIC) and added that "I honestly believe that we will make a fool of ourselves with this ruling. Europe has launched a negative alert against any biotechnological company intending to establish and develop new consumption varieties thanks to gene editing." Dozens of statements from academies and research institutes all over the world were issued against the ruling of the Court of Justice of

The truth is that the speed with which new technologies are generated, the difficulties inherent in establishing a distinction between conventional and biotechnological genetic breeding, together with the capacity to identify and differentiate their products, results in that the credibility and feasibility of biosafety approaches based both on processes and products are under a lot of criticism (Marchant and Stevens 2015; NAP 2016).

In a reasonable scenario, everything seems to indicate that the biosafety framework of "Agriculture 3.0" applied on GMOs would have no scientific support to be applied on products resulting from "Agriculture 4.0" NBTs. For example, the recent revised arguments that led to the development of current regulatory frameworks have permitted to realize that the size and extent of a genetic change in itself (whether via conventional breeding or by application of modern technology) have relatively little importance in connection with the degree of change in a plant and, therefore, the risk presented to the environment or the harmlessness of food. The need to coordinate efforts among countries and to ensure that regulatory frameworks be based on science has been ratified by many countries (WTO 2018).[7] However, the ruling of the CJEU is inconsistent with this aspiration.

Finally, as explained before and with few exceptions, only large biotechnology-based seed companies have actively participated in "Agriculture 3.0," establishing a resulting relationship between patents and regulation. While a new biotechnological development does not necessarily mean significant progress or a change in the scientific paradigm (as a matter of fact, most developments were not), the point is that the only way to recover such transgenic biotechnology investments is via strong intellectual property protection. In "Agriculture 3.0," intellectual property and regulatory framework became entangled in a cause–effect relationship.

This leads to a substantial question: if the new technologies of "Agriculture 4.0" attained significant productive progress, but at the same time these techniques were affordable, easier to apply, and were left outside the biosafety regulatory framework, would derived products have a weak intellectual property protection framework or even not have any at all? What would be the logics of this argumentation?

Multiple studies state that IPRs must always be related to the degree of industrialization and that the step from imitation to innovation must be done accompanying that development with stronger and more effective systems (Chen and Puttitanuna 2005). Therefore, it is not innovation complexity, but the level of development that defines the optimal level of protection via IPRs (Correa 2016).

---

the European Union, and a specific mention is due to the position paper signed by 85 European scientists on October 24, 2018: http://www.vib.be/en/news/Documents/Position%20paper%20on%20the%20ECJ%20ruling%20on%20CRISPR%2024%20Oct%202018.pdf.

[7] On October 26, 2018 the World Trade Organization received a joint communication executed by Argentina, Australia, Brazil, Canada, the United States, Guatemala, Honduras, Paraguay, the Dominican Republic, and Uruguay, consisting in an "International Statement on Agricultural Applications of Precision Biotechnology."

2.3 Five Examples for Analysis

Let us now analyze the following five examples:

1. Plants tolerating herbicides, especially glyphosate, are one of the two most commercially successful GMOs. In 2016, Cibus scientists used a combination of Oligonucleotide-Directed Mutagenesis (ODM) and CRISPR-Cas9 to develop a flax variety *(Linum usitatissimum)* accurately editing the genes of the 5′-enolpyruvylshikimate-3-phosphate synthase (EPSPS) enzyme. This enzyme is part of the biosynthetic route of aromatic amino acids and is directly responsible for the sensitivity of all plants against the glyphosate herbicide. The resulting regenerated plants were tolerant to glyphosate herbicide, also showing a Mendelian inheritance of the edited character (Sauer et al. 2016). The difference with the plant varieties that tolerated glyphosate of all crops known so far is that this flax variety is not transgenic because, strictly speaking, no gene of a similar or different species has been introduced in it. The locus of enzyme EPSPS was edited changing two bases of cytokine to one guanine and one tyrosine.

   - The capacity to specifically edit a gene requires deep knowledge of the target species' genome, including the very location of the gene to the accurate identification of nucleic bases that require modification. The EPSPS gene of flax, which was modified to make it tolerant to the glyphosate herbicide, exactly emulated the same achievement that at its time was the introgression of bacteria gene into a higher plant and the resulting development of genetically modified plants that were tolerant to such herbicide. But this edition did not result in a transgenic plant, and it should not be regulated as such either. Now, the natural gene was modified and, from an intellectual property point of view, there are two problems:

      (a) Would a patent be issued for a natural gene for which a pair of bases was modified? Identifying a cutoff point and establishing that 1, 2, 3, or *n* changes of bases represent an inventive step meeting such patentability criterion has no scientific support at all. A single change of bases could be highly inventive in a case, and *n* changes could be insignificant in another.
      (b) Would breeder rights be recognized to the new, genetically edited, and glyphosate-tolerant flax variety? If this new edited variety that has not been genetically modified is an isoline of an alleged prior genetically modified variety on the same expression character, both varieties would be undistinguishable from a phenotype point of view and would, therefore, be identical. It would not be possible to protect the new flax variety with breeder rights under the UPOV Convention.

2. In corn, the ARGOS8 gene is a negative regulator of the response to ethylene. Plants with overexposure to ARGOS8 show reduced sensitivity to ethylene and improve the production of grain under water stress conditions. In 2016, Pioneer scientists used CRSPR-Cas9 to edit native corn promoter gene GOS2, which

grants a moderate constitutive level of expression, and to insert it in a non-translated region of native gene ARGOS8 to replace the native promoter of this gene. Regenerated plants showed high expression of ARGOS8, which was detectable in all plant tissues, and in field essays they significantly increased their grain production under water-stress conditions in blooming, without showing at the same time any performance decrease under adequate humidity conditions (Shi et al. 2016). New varieties are not transgenic.

- The edition of gene ARGOS8's promoter in corn and the insertion of it in a non-translated region with the purpose of increasing its expression and obtaining water-stress-resistant plants is an exceptional achievement. As in the case of flax, this required deep knowledge of the species' genome and high technical capacity to attain the intended edition. This could be done with conventional crossing procedures if there is a compatible organism in which there is the change sought and subsequent selection by backcrossing or transgenesis. The edition of this gene's promoter did not turn this corn variety into transgenic.

  (a)Would a patent be issued for the edited promoter of gene ARGOS8? It is likely that patent office considers that the "new" promoter is still the natural and preexisting promoter even if edited.
  (b) Would breeder rights be recognized on the new drought-tolerant corn variety? If this new corn variety with an edited promoter gene and with no genetic modification were an isoline of an alleged prior genetically modified variety in the same expression character, both varieties would be phenotypically indistinguishable and, therefore, identical, even if the edited variety is significantly tolerant to environments under water stress. It would not be possible to protect the new corn variety or its component lines with breeder rights under the UPOV Convention.

3. The standard corn grain's endosperm is mainly made up of two types of molecules: amylose and amylopectin. Gene Wx1 codifies for the starch synthase enzyme, key for amylose synthesis. There is a natural mutation of this gene, which is wx1-waxy, and has been marketed as a specialty for the industry for decades as endosperm is only made up by amylopectin. Pioneer scientists successfully suppressed the expression of gene Wx1 via the use of technique CRISP-Cas9. The new product was presented to the US regulators and they decided that this new edited corn is not transgenic, and as such became the first product officially out of the regulatory procedure in that country (Waltz 2016; USDA 2016).

- One of the most successful uses of editing techniques such as CRISPR-Cas9 is in attaining a genetic "knockout," that is, targeting the edition of a specific gene of a biosynthetic route so as to deactivate its function. This is the case of suppressing the expression of gene Wx1 responsible for the synthesis of amylose in corn. But in this very example there is a mutation of this gene, wx1-waxy, already existing in nature and producing the same effect. Then, if we wanted to obtain a corn line without producing amylose, it would be possible to use the natural mutant gene, crossing, and subsequent selection, or to deactivate the amylose-synthesis gene via gene "knockout."

- Over what could a patent be issued? In this case, the problem is the inexistence of the patentable "object," as the edition entailed the elimination of such object.
- Would breeder rights be recognized on the new edited, not genetically edited waxy corn variety? If this new corn variety has been previously obtained with conventional crossings and subsequent selection based on crossing with a mutant waxy line, both varieties would be undistinguishable phenotypically and, therefore, identical. It would not be possible to protect the new corn variety or its component lines with breeder rights under the UPOV Convention.

4. PGRs, as mentioned in Chap. 1, are that part of genetic diversity with current or potential value and are found in the wild, at times protected by local communities, at times collected and catalogued, and they are found in germplasm banks, whether on-site or off-site. The creation and new generation of variants in gene bottoms of species is the main requirement of any gene-breeding program. There are many existing and properly managed germplasm banks, but most species do not have them. Rani et al. (2016) argue that gene-editing techniques have a potential role in the creation of variation in germplasm so as to become a key part in the increase and generation of variability, which would be key in species without germplasm banks or with narrow variability. However, nothing prevents these techniques being applied in any species with extensive PGRs as yet not exploited.

• Tomato comes from the South American Andes, from the center of Ecuador, to Peru, and northern Chile, including the Galapagos Islands. Precisely in these islands in 1950, a group of researchers found a species of wild tomato without joints between the central stem and the fruits, so fruits did not detach early and remained tied to the plant for longer, which is extremely important for mechanic harvesting. When the "jointless" feature was introduced in the cultivated varieties of tomatoes, it became evident that such feature was connected with another characteristic whereby branches with flowers produced many additional branches with more flowers, which resulted in very small fruits—a commercially undesirable effect. A team of researchers led by Zachary Lippman from the Cold Spring Harbor Laboratory in the United States analyzed a collection of 4193 wild tomato varieties, seeking varieties with unusual ramification patterns, identifying the two genes with the characteristics mentioned (Lippman et al. 2008). After discovering these genes, the team used the gene edition of CRISPR-Cas9 to eliminate its activity in a commercial tomato variety, as well as that of a third gene that also affects the number of flowers, in multiple combinations. This resulted in a gamut of architectures in the plant, from long and thin branches with flower bunches, to thick little flower bunches with an architecture similar to that of cauliflower, including some with improved performance (Soyk et al. 2017). In other words, for this development, the material that forms the genetic resource has not been used and no physical access to it was taken, just its information.

- Over what could a patent be issued? The problem here is the lack of novelty. If the edition of the gene of the commercial variety was to imitate—base-by-base—the structure of the genetic resource gene, there would be no invention, even if this is a technical challenge at the border of scientific knowledge and a huge technical breakthrough.
- Would breeder rights be recognized on the tomato variety with modified plant architecture? If the rights were on already existing varieties, most likely yes. However, as the genes responsible for the features are known, the same result could be obtained via traditional breeding (albeit in a much longer period), and would be phenotypically indistinguishable and, therefore, identical.
- Was a genetic resource used and should shared benefits be recognized to the countries where such resources originated? Technically speaking and considering the current language of treaties, in this case there has been no access to the genetic resource and there would be no obligation to compensate the country that provided the resource.

5. The "cotton bollworm," also known as "Old World bollworm" (*Helicoverpa armigera*) is one of the most devastating pests all over the world on cotton crops. Serious damage because of this pest has been reported in Europe, Asia, Africa, Oceania, and South American. In South America, the pest was identified in 2011 in Brazil and in 2013 in Argentina. A team of Southwest University researchers in China led by Ming-Hui Jin successfully introduced predictable mutations and at the same time generated several types of genetic modifications using combinations of pairs of sgRNA and Cas9, including mutations in genes HaCad and HaABCC2, in this worm's genome using the CRISPR-Cas technique. This is how these researchers could deactivate the action of the pest on cotton (Ming-Hui et al. 2018).

   • The fight against pests in crops is one of the most important specialties in agriculture. But strategies have always been focused on developing defense mechanisms in the genome of the plants of the crop itself. This Chinese research team did the opposite. They altered the pest's genome so that it incorporated a modification whereby its effect on crops was deactivated.

   - Over what could a patent be issued or a breeder right recognized? In this case, it is clear that nothing pertaining to the crop, as none of its elements have been modified. The significant progress for the protection of the crop was made modifying the pest's genome and it is more than difficult to imagine that any property rights may be applied. Simply put, this case has no solution based on the existing intellectual property mechanisms.

The theory in this chapter and these five actual and specific examples show part of the magnitude of the problem—products resulting from NBTs pertain to

"Agriculture 4.0." When regulators pretend to follow the rules of "Agriculture 3.0," they are showing the inconsistencies of the TRIPS and the CBD, and the inconsistencies of an important part of the specific legislation that is now outdated by technical progress.

Last, there is yet another element contributing to this complex scenario. In most countries, granting patents, breeder rights over plant varieties, PGRs, and biosafety are the province of multiple offices, which in turn are under a separate department or ministry. There is generally no contact between offices, or any potential contact is limited to the formation of special groups to discuss a specific issue.

References

Abudayyeh OO, Gootenberg JS, Essletzbichler P, Han S, Joung J, Belanto JJ, Verdine V, Cox DBT, Kellner MJ, Regev A, Lander ES, Voytas DF, Ting AY, Zhang F (2017) RNA targeting with CRISPR-Cas13. Nature 550(7675):280–284

Alkan F, Wenzel A, Anthon C, Havgaard JH, Gorodkin J (2018) CRISPR-Cas9 off-targeting assessment with nucleic acid duplex energy parameters. Genome Biol 19:177. https://doi.org/10.1186/s13059-018-1534-x

APHIS (2017) Importation, interstate movement, and environmental release of certain genetically engineered organisms. A proposed rule by the Animal and Plant Health Inspection. Service on 01/19/2017. https://www.federalregister.gov/documents/2017/01/19/2017-00858/importation-interstate-movement-and-environmental-release-of-certain-genetically-engineered

Barraclough E (2013) What Myriad means for biotech. WIPO Mag 4/2013

Bass K (2015) The battle over plant genetic resources: interpreting the international treaty for plant genetic resources. Chic J Int Law 16(1):7. Available at: http://chicagounbound.uchicago.edu/cjil/vol16/iss1/7

Bergadá P, Rapela M, Enríquez R, Risso D, Mendizabal J (2016) Generating value in the soybean chain through royalty collection: an international study. BioSci Law Rev 15(5):169–210

Bhatti S (2016) Use it or lose it: the international treaty provides access to key plant breeding material. Eur Seed 3(4):19–22

Boyle J (2003) The second enclosure movement and the construction of the public domain. Law Contemp Probl 66:33–74. http://creativecommons.org/licenses/by-sa/1.0. It is also available at http://www.law.duke.edu/journals/66LCPBoyle

Bragdon S (ed) (2004) International law of relevance to plant genetic resources: a practical review for scientists and other professionals working with plant genetic resources. Issues in genetic resources, No 10, March 2004. International Plant Genetic Resources Institute, Rome

Campi M (2016a) The effect of intellectual property rights on agricultural productivity. Agric Econ 48:1–13

Campi M (2016b) Innovation and intellectual property rights: the case of soybean seeds in Argentina and the United States. In: Al-Hakim L, Wu X, Koronios A, Shou Y (eds) Handbook of research on driving competitive advantage through sustainable, lean, and disruptive innovation. Pennsylvania IGI Global, Hershey, pp 334–354

CAST (2018a) Regulatory barriers to the development of innovative agricultural biotechnology by small businesses and universities. CAST – Council for Agricultural Science and Technology, March 2018, Number 59

CAST (2018b) Genome editing in agriculture: methods, applications, and governance. CAST – Council for Agricultural Science and Technology, July 2018, Number 60

CBD (2016) Convention on biological diversity, subsidiary body on scientific, technical and technological advice. Twentieth meeting Montreal, Canada, 25–30 April 2016 Agenda item 6 Synthetic Biology, Draft recommendation submitted by the Chair

CBD (2018a) Synthetic biology. Draft decision submitted by the Chair. SBSTTA, Subsidiary Body on Scientific, Technical and Technological Advice. Twenty-second meeting Montreal, Canada, 2–7 July 2018. CDB/SBSTTA/22/L.6

CBD (2018b) Digital sequence information on genetic resources. Draft decision submitted by the Chair. SBSTTA, Subsidiary Body on Scientific, Technical and Technological Advice. Twenty-second meeting Montreal, Canada, 2–7 July 2018. CDB/SBSTTA/22/CRP.10

Ceasar SA, Rajan V, Prykhozhij SV, Berman JN, Ignacimuthu S (2016) Insert, remove or replace: a highly advanced genome editing system using CRISPR/Cas9. Biochim Biophys Acta 1863:2333–2344

Chen A, Puttitanuna T (2005) Intellectual property rights and innovation in developing countries. J Dev Econ 78:474–493

Cong L, Ran FA, Cox D, Lin S, Barretto R, Habib N, Hsu PD, Wu X, Jiang W, Marraffini LA, Zhang F (2013) Multiplex genome engineering using CRISPR/Cas systems. Science 339:819–823. https://doi.org/10.1126/science.1231143

Correa CM, Shashikant S, Meienberg F (2015) Plant variety protection in developing countries: a tool for designing a sui generis plant variety protection system: an alternative to UPOV 1991. Association for Plant Breeding for the Benefit of Society (APBREBES) and its member organizations: Berne Declaration, Development Fund, SEARICE, Third World Network. Available at http://www.apbrebes.org/news/new-publication-plant-variety-protection-developing-countries-tool-designing-sui-generis-plant. Accessed 24 Oct 2017

Correa CM (2016) IP and economic development. In: WIPO conference on intellectual property and economic development, Geneva, 7 Apr to 8 Apr 2016

CTNBio (2018) Resolução Normativa N° 16, de 15 de janeiro de 2018

de Felipe M, Gerde JA, Rotundo JL (2016) Soybean genetic gain in maturity groups III to V in Argentina from 1980 to 2015. Crop Sci 56(6). https://doi.org/10.2135/cropsci2016.04.0214

Duensing N, Sprink T, Parrott WA, Fedorova M, Lema MA, Wolt JD, Bartsch D (2018) Novel features and considerations for ERA and regulation of crops produced by genome editing. Front Bioeng Biotechnol 6:1–16

Dugar G, Leenay RT, Eisenbart SK, Bischler T, Aul BU, Beisel CL, Sharma CM (2018) CRISPR RNA-dependent binding and cleavage of endogenous RNAs by the Campylobacter jejuni Cas9. Mol Cell 69(5):893–905.e7. https://doi.org/10.1016/j.molcel.2018.01.032

Dutfield GM, Roberts TW (2017) Intellectual property rights. In: Thomas B, Murray BG, Murphy D (eds) Encyclopedia of applied plant sciences, Breeding genetics and biotechnology, vol 2, 2nd edn. Elsevier, Academic Press, Amsterdam, pp 23–27

Egelie KJ, Graff GD, Strand SP, Johansen B (2016) The emerging patent landscape of CRISPR-Cas gene editing technology. Nat Biotechnol 34(10):1025–1031

European Academies (2015) New breeding techniques. European Academies Science Advisory Council. http://www.interacademies.net/File.aspx?id=28130

European Commission (2017) New plant breeding techniques. http://ec.europa.eu/food/plant/gmo/legislation/plant_breeding_en

Eyhérabide G (2015) Evolución y cambios recientes de los rendimientos medios nacionales del cultivo de maíz en Argentina. Revista de Tecnología Agropecuaria INTA Pergamino 10(29):6–12

Falck-Zepeda J, Wesseler J, Smyth SJ (2013) The current status of the debate on socioeconomic regulatory assessments: positions and policies in Canada, the USA, the EU and developing countries. World Rev Sci Technol Sust Dev 10:203–227. https://doi.org/10.1504/WRSTSD.2013.057690

FDA (2017) Genome editing in new plant varieties used for foods; request for comments. A notice by the Food and Drug Administration on 01/19/2017. https://www.federalregister.gov/documents/2017/01/19/2017-00840/guidance-genome-editing-in-new-plant-varieties-used-for-foods

Gao C (2018) The future of CRISPR technologies in agriculture. Nat Rev Mol Cell Biol 19:275–276

Genetic Literacy Project (2016) How are governments regulating CRISPR and New Breeding Technologies (NBTs)? http://gmo.geneticliteracyproject.org/FAQ/how-are-governments-regulating-crispr-and-new-breeding-technologies-nbts/

Georges F, Ray H (2017) Genome editing of crops: A renewed opportunity for food security. GM Crops & Food 8:1–12

GHR (2018) What are genome editing and CRISPR-Cas9. Genetics Home Reference. https://ghr.nlm.nih.gov/primer/genomicresearch/genomeediting

Greiber T, Peña Moreno S, Åhrén M, Nieto Carrasco J, Kamau EC, Cabrera Medaglia J, Oliva MJ, Perron-Welch F, in cooperation with Ali N, Williams C (2012) An explanatory guide to the Nagoya Protocol on access and benefit-sharing, vol XVIII. IUCN, Gland, 399 pp

Grens K (2018) Japanese authorities recommend not regulating gene editing. The Scientist, Aug 2018. https://www.the-scientist.com/news-opinion/japanese-authorities-recommend-not-regulating-gene-editing-64675

Harrington LB, Burstein D, Chen JS, Paez-Espino D, Ma E, Witte IP, Cofsky JC, Kyrpides NC, Banfield JF, Doudna JA (2018) Programmed DNA destruction by miniature CRISPR-Cas14 enzymes. Science 2018:eaav4294. https://doi.org/10.1126/science.aav4294

Hein T (2018) No access, no benefits – part 3 – the view from academia. European Seed, posted on December 12th, 2018 by Treena Hein. Int News Regul 5(4)

ICA (2018) Resolución del Instituto Colombiano Agropecuario No 29.299 del 01/08/18

ICC (2017) Summary of Convention on Biological Diversity COP-13 and Nagoya Protocol meetings. Commission on Intellectual Property, International Chamber of Commerce, 9 Jan 2017 DYE/abs

ISF (2003) Disclosure of origin in intellectual property protection applications. Position paper of the International Seed Federation, adopted in Bangalore, June 2003

ISF (2012) ISF view on Intellectual Property. Position paper of the international seed federation, adopted in Rio de Janeiro, Brazil, 28 June 2012

ISF (2018) Digital sequence information. Position paper of the international seed federation. June 2018

Ishii T, Araki M (2016) A future scenario of the global regulatory landscape regarding genome-edited crops. GM Crops Food:22–34. Published online: 14 Dec 2016. https://doi.org/10.1080/21645698.2016.1261787

Israel (2017) Ministry of Agriculture and Rural Development, National Committee for Transgenic Plants (NCTP) summary of NCTP meeting of 08/08/16. Published on 5 Mar 2017

Jiang W, Bikard D, Cox D, Zhang F, Marraffini L (2013) RNA-guided editing of bacterial genomes using CRISPR-Cas systems. Nat Biotechnol 31:233–239

Jinek M, Chylinski K, Fonfara I, Hauer M, Doudna JA, Charpentier E (2012) A programmable dual-RNA-guided DNA endonuclease in adaptive bacterial immunity. Science 337:816–821. https://doi.org/10.1126/science.1225829

Jones HD (2015) Future of breeding by genome editing is in the hands of regulators. GM Crops Food 6(4):223–232. https://doi.org/10.1080/21645698.2015.1134405

Jones HD (2016) Are plants engineered with CRISPR technology genetically modified organisms? Biochem Soc, June 2016:14–17

Kariyawasam K, Tsai M (2018) Access to genetic resources and benefit sharing – implications of Nagoya Protocol on providers and users. J World Intellect Prop 21(5–6):289–305

Kim H, Kim S-T, Ryu J, Kang B-C, Kim J-S, Kim S-G (2017) CRISPR/Cpf1-mediated DNA-free plant genome editing. Nat Commun. https://doi.org/10.1038/ncomms14406

Kistler L, Yoshi Maezumi S, de Souza JG, Przelomska NAS, Malaquias Costa F, Smith O, Loiselle H, Ramos-Madrigal J, Wales N, Rivail Ribeiro E, Morrison RR, Grimaldo C, Prous AP, Arriaza B, Gilbert MTP, de Oliveira Freitas F, Allaby RG (2018) Multiproxy evidence highlights a complex evolutionary legacy of maize in South America. Science 362(6420):1309. https://doi.org/10.1126/science.aav0207

Kock M (2009) Patents for life: the role of intellectual property rights on plant innovations. BioSci Law Rev 10(5):167–176

Kock M (2013) Adapting IP to an evolving agricultural innovation landscape. WIPO Mag. http://www.wipo.int/wipo_magazine/en/2013/02/article_0007.html

Kozubek L (2016) Modern Prometheus: editing the human genome with CRISPR-Cas9. Cambridge University Press

Kwong M (2014). Six significant moments in patent history. Oil in the Spotlight, 4 Nov 2014.

Lange CE, Federizzi LC (2009) Estimation of soybean genetic progress in the south of Brazil using multienvironmental yield trials. Sci Agric (Piracicaba, Braz) 66(3):309–316

Liang Z, Chen K, Li T, Zhang Y, Wang Y, Zhao Q, Liu J, Zhang H, Liu C, Ran Y, Gao C (2017) Efficient DNA-free genome editing of bread wheat using CRISPR/Cas9 ribonucleoprotein complexes. Nat Commun 8:14261. https://doi.org/10.1038/ncomms14261

Lippman ZB, Cohen O, Alvarez JP, Abu-Abied M, Pekker I, Paran I et al (2008) The making of a compound inflorescence in tomato and related nightshades. PLoS Biol 6(11):e288. https://doi.org/10.1371/journal.pbio.0060288

Lusser M, Parisi C, Plan D, Rodríguez-Cerezo E (2011) New plant breeding techniques – state-of-the-art and prospects for commercial development. European Commission's Joint Research Centre (JRC), Institute for Prospective Technological Studies (IPTS), JRC Institute for Health and Consumer Protection (IHCP) EUR 24760 EN – 2011

Mali P, Yang L, Esvelt KM, Aach J, Guell M, DiCarlo JE, Norville JE, Church GM (2013) RNA-guided human genome engineering via Cas9. Science 339:823–826. https://doi.org/10.1126/science.1232033

Marchant GE, Stevens YA (2015) A new window of opportunity to reject process-based biotech-nology regulation. GM Crops Food 6:233–242

McDougall P (2011) The cost and time involved in the discovery, development and authorization of a new plant biotechnology derived trait. A consultancy study for CropLife International

McHughen A (2012) Introduction to the GM crops special issue on biosafety, food and GM regula-tion. GM Crops Food 3(1):6–8. https://doi.org/10.4161/gmcr.17646

Ming-Hui J, Yu-Tao X, Ying C, Jie H, Chao-Bin X, Kong-Ming W (2018) Chromosomal deletions mediated by CRISPR/Cas9 in Helicoverpa armigera. Insect Sci. https://doi.org/10.1111/1744-7917.12570

Mojica JM, Montoliu L (2016) On the origin of CRISPR-Cas technology: from prokaryotes to mammals. Trends Microbiol 24(10):811–820

Moore G, Tymowski W (2005) Explanatory guide on the international treaty on plant genetic resources for food and agriculture. IUCN, Gland and Cambridge

Myszczuk AP, De Meirelles ML (2014) Patents and living matter: the construction of a patent sys-tem attractive to biotechnology. In: de Miguel Beriain I, Romeo Casabona CM (eds) Symbio and human health: a challenge to the current IP framework? Springer Science+Business Media, Dordrecht

NAP (2016) Genetically engineered crops: experience and prospects.. Committee on Genetically Engineered Crops: past experience and future prospects; Board on Agriculture and Natural Resources; Division on Earth and Life Studies; National Academies of Sciences, Engineering, and Medicine. The National Academies Press, Washington, DC. https://doi.org/10.17226/23395.

NAP (2017) Advancing concepts and models for measuring innovation: proceedings of a work-shop. National Academies of Sciences, Engineering, and Medicine. The National Academies Press, Washington, DC. https://doi.org/10.17226/23640

Nature Editorials (2017) Legal Limbo: Europe is drugging its feet on gene editing rules and scien-tists should push the issue. Nature 242:392

Noriega IL, Halewood M, Galluzzi G, Vernooy R, Bertacchini E, Gauchan D, Welch E (2013) How policies affect the use of plant genetic resources: the experience of the CGIAR. Resources 2:231–269

OJEU (2017) Request for a preliminary ruling from the Conseil d'État (France) lodged on 17 October 2016 — Confédération paysanne, Réseau Semences Paysannes, Les Amis de la Terre France, Collectif vigilance OGM et Pesticides 16, Vigilance OG2M, CSFV 49, OGM: dangers, Vigilance OGM 33, Fédération Nature et Progrès v Premier ministre, Ministre de l'agriculture, de l'agroalimentaire et de la forêt (Case C-528/16) (2017/C 014/29) Language of the case: French. Off J Eur Union (OJEU), 16.01.2017

OJEU (2018a) Opinion of Advocate General Bobej, delivered on 18 January 2018. Case C-528/16. Off J Eur Union (OJEU), 18-01-2018

OJEU (2018b) Judgment of the court in case C-528/16. Off J Eur Union (OJEU) 25-07-2018

Overdijk TFW (2013) Essentially derived varieties: case law in the Netherlands and connected observations. UPOV EDV SEMINAR Geneva, 22 October 2013

Overmann J, Scholz AH (2017) Microbiological research under the Nagoya Protocol: facts and fiction. Sci Soc 25(2):85–88. https://doi.org/10.1016/j.tim.2016.11.001

Phillips PWB (2017) Ownership of plant genetic resources. In: Thomas B, Murray BG, Murphy D (eds) Encyclopedia of applied plant sciences, Breeding genetics and biotechnology, vol 2, 2nd edn. Elsevier, Academic Press, Amsterdam, p 28

Prathapan D, Pethiyagoda R, Bawa KS, Raven PH, Rajan PD, 172 co-signatories from 35 countries (2018) When the cure kills: CBD limits biodiversity research. Science 360(6396):1405–1406

Puchta H (2017) Applying CRISPR/Cas for genome engineering in plants: the best is yet to come. Curr Opin Plant Biol 36:1–8

Rani R, Yadav P, Barbadikar KM, Baliyan N, Malhotra EV, Singh BK, Kumat A, Singh D (2016) CRISPR/Cas9: a promising way to exploit genetic variation in plants. Biotechnol Lett 38:1991–2006

Rapela MA (2000) Derechos de propiedad intelectual en vegetales superiores. Editorial Ciudad Argentina, 466 páginas

Rapela MA (2005) Plantas transgénicas, bioseguridad y principio precautorio. Editorial de la Universidad Nacional de La Plata, 570 páginas

Rapela MA (2006) Excepción del fitomejorador: de la libre disponibilidad a la variedad esencialmente derivada. En: "Innovación y Propiedad Intelectual en Mejoramiento Vegetal y Biotecnología Agrícola", Rapela, Miguel Ángel, (Director Académico), Gustavo J. Schötz (coordinador), Enrique del Acebo Ibáñez, Juan Miguel Massot, Helena María Noir, Fernando Sánchez, Andrés Sánchez Herrero, María Celina Strubbia y Mónica Witthaus. Editorial Heliasta, páginas 207–242

Rapela MA (2008) El concepto de Variedad Esencialmente Derivada y la Excepción al Fitomejorador dentro del Derecho del Obtentor. 2° Congreso Nacional e Internacional de Agrobiotecnología, Propiedad Intelectual y Políticas Públicas. Universidad Nacional de Córdoba, 27 a 29 de agosto de 2008

Rapela MA (2010) Farmer's exception, farmer's rights and other related issues. Seed News XIV(1):28–29

Rapela MA (2013) Patents and the seed industry. Seed News XVII(3):20–25

Rapela MA (2014a) La era post transgénica y el desafío de las nuevas técnicas de mejoramiento. Actas del Seminario organizado por el Instituto de Genética "Ewald Favret" del INTA Castelar en conmemoración del 45 Aniversario de la Sociedad Argentina de Genética y los 50 años de la creación del híbrido de maíz forrajero. Castelar, 5 de diciembre 2014

Rapela MA (2014b) Post-Transgenesis: new plant breeding techniques. Seed News Mag XVIII:14–15

Rapela MA (2014c) The Nagoya protocol. Seed News XVIII(6):16–19

Rapela MA (2015) The adoption of conventions and treaties related to genetic resources and intellectual property issues: current situation and status in the SAA region. Conference at the 5th congress of the Seed Association of the Americas. Cancún, México, 10 Sept 2015

Rapela MA (2016) Ley 20.247 de Semillas y Creaciones Fitogenéticas: las razones para su actualización y los proyectos bajo análisis en Argentina. Revista Interdisciplinaria de Estudios Agrarios, Facultad de Ciencias Económicas, Universidad de Buenos Aires, No 43, 2° Semestre 2016. No 45: 69–98

Rapela MA (2018a) Gene editing and CRISPR-Cas. Seed News Mag XXII:12–16

Rapela MA (2018b) Metodología de CRISPR, aspectos legales y regulatorios. Actas XI Congreso Nacional de Maíz, Mesa de Genética y Mejoramiento Genético Vegetal, págs. 266–270

Rapela MA (2018c) Edición Génica mediante sistemas CRISPR/Cas. AGROPOST CPIA-Consejo Profesional de Ingeniería Agronómica, No 155, abril-mayo, págs 11–13

Rapela MA, Levitus G (2014) Novas técnicas do melhoramento. In: Anuario da ABRASEM. Associação Brasileira de Sementes e Mudas, Páginas, pp 29–32

Rojas B, Lorena D (2013) Vicissitudes of Nagoya Protocol in Colombia. 16(3):17–23
Ruiz Muller M (2015) Genetic resources as natural information: implications for the convention on biological diversity and Nagota Protocol. Taylor & Francis Ltd, New York, 170 pp
Ruiz Muller M, Caillaux Zazzali J (2014) Propiedad Intelectual y acceso a Recursos Gnéticos en un ambiente altamente politizado y "desinformado". Anuario Andino de Derechos Intelectuales X(10):317–332
Ruiz Muller M, Henry Vogel J, Zamudio T (2010) La lógica debe prevalecer: un nuevo marco teórico y operativo para el Régimen Internacional de Acceso a RGV y Distribución Justa y Equitativa de Beneficios. Documentos de Investigación V(13)
SAGYP (2015) Resolución 173/2015 estableciendo procedimientos de los productos derivados de nuevas técnicas de mejoramiento. Secretaría de Agricultura, Ganadería y Pesca de Argentina. http://servicios.infoleg.gob.ar/infolegInternet/anexos/245000-249999/246978/norma.htm
Samanta MK, Dey S, Gayem S (2016) CRISPR/Cas9: an advanced tool for editing plant genomes. Transgenic Res 25:561–573
Sauer NJ, Narváez-Vásquez J, Mozoruk J, Miller RB, Warburg ZJ, Woodward MJ, Mihiret YA, Lincoln TA, Segami RE, Sanders SL, Walker KA, Beetham PR, Schöpke CR, Gocal GFW (2016) Oligonucleotide-mediated genome editing provides precision and function to engineered nucleases and antibiotics in plants. Plant Physiol 170(4):1917–1928. https://doi.org/10.1104/pp.15.01696
Scheben A, Wolter F, Batley J, Puchta H, Edwards D (2017) Towards CRISPR/Cas crops – bringing together genomics and genome editing. New Phytol 216:682. https://doi.org/10.1111/nph.14702
Schmidt S (2018) To regulate or not to regulate: current legal status for gene-edited crops. Global Engage. http://www.global-engage.com/agricultural-biotechnology/to-regulate-or-not-to-regulate-current-legal-status-for-gene-edited-crops/
Schiml S, Puchta H (2016) Revolutionizing plant biology: multiple ways of genome engineering by CRISPR/Cas. Plant Methods 12:8. https://doi.org/10.1186/s13007-016-0103-0
Schindele P, Wolter F, Puchta H (2018) Transforming plant biology and breeding with CRISPR/Cas9, Cas12 and Cas13. FEBS Lett:13073. https://doi.org/10.1002/1873-3468
Schuttelaar & Partners (2016) The regulatory status of new breeding techniques in countries outside the European Union. Document developed by Schuttelaar & Partners Version. Schuttelaar & Partners, The Hague, The Netherlands, June 2015; 65 pp
Shi J, Gao H, Wang H, Lafitte HR, Archibald RL, Yang M, Hakimi SM, Mo H, Habben JE (2016) ARGOS8 variants generated by CRISPR-Cas9 improve maize grain yield under field drought stress conditions. Plant Biotechnol J 15:207. https://doi.org/10.1111/pbi.12603
Shimatani Z, Fujikura U, Ishii H, Matsui Y, Suzuki M, Ueke Y, Taoka K, Terada R, Nishida K, Kondo A (2018) Inheritance of co-edited genes by CRISPR-based targeted nucleotide substitutions in rice. Plant Physiol Biochem 131:78. https://doi.org/10.1016/j.plaphy.2018.04.028
Shreya, Rana K, Ainmisha (2017) CRISPR/Cas9: a nobel approach for genome editing. Int J Curr Microbiol App Sci 6(5):1866–1871. https://doi.org/10.20546/ijcmas.2017.605.205
SINGER (2016) Data base of the System-Wide Information Network for Genetic resources (SINGER). http://www.singer.cgiar.org. Accessed Sept 2016
Smith JSC, Jones ES, Nelson BK (2013) The use of molecular marker data to assist in the determination of essentially derived varieties. In: Tuberosa R, Graner A, Frison E (eds) Genomics of plant genetic resources. Vol 1. Managing, sequencing and mining genetic resources. Springer, Dordrecht, pp 49–66
Smyth SJ, Phillips PWB (2014) Risk, regulation and biotechnology: the case of GM crops. GM Crops & Food 5(3):170–177. https://doi.org/10.4161/21645698.2014.945880
Smyth SJ, McDonald J, Falck-Zepeda JB (2014) Investment, regulation, and uncertainty: Managing new plant breeding techniques. GM Crops Food Biotechnol Agric Food Chain 5:4–3. https://doi.org/10.4161/gmcr.27465
Soyk S, Lemmon ZH, Oved M, Fisher J, Liberatore KL, Park SJ, Goren A, Jiang K, Ramos A, van der Knaap E, van Eck J, Zamir D, Eshed Y, Lippman ZB (2017) Bypassing negative epis-

tasis on yield in tomato imposed by a domestication gene. Cell. https://doi.org/10.1016/j.
cell.2017.04.032

Tagliabue G (2016) The EU legislation on "GMOs" between nonsense and protectionism: an ongoing Schumpeterian chain of public choices. GM Crops Food:35–51. https://doi.org/10.10 80/21645698.2016.1270488

Urnov FD (2018) Genome editing B.C. (before CRISPR): lasting lessons from the "old testament". CRISPR J 1:34–46

USDA   (2016).   https://www.pioneer.com/CMRoot/Pioneer/About_Global/Non_ Searchable/15-352-01_air_response_signed.pdf

USDA (2018) Secretary Perdue Issues USDA statement on plant breeding innovation. USDA Animal and Plant Health Inspection Service, Washington, D.C., 28 Mar 2018

Van den Hurk A (2011) Access to genetic resources for vegetable breeding. International Seed Federation, World Seed Congress, Belfast, 31 May 2011

Waltz E (2016) CRISPR-edited crops free to enter market, skip regulation. Nat Biotechnol 34:582– 582. https://doi.org/10.1038/nbt0616-582

Whelan AI, Lema MA (2015) Regulatory framework for gene editing and other new breeding techniques (NBTs) in Argentina. GM Crops& Food 6(4):253–265

Whelan AI, Lema MA (2017) A research program for the socioeconomic impacts of gene editing regulation. GM Crops Food 8:52–61. https://doi.org/10.1080/21645698.2016.1271856

WIPO (2017) What is a Patent? World Intellectual Property Organization. http://www.wipo.int/ patents/en/. Accessed Jan 2017

WIPO (2018a) WIPO Intergovernmental Committee on Intellectual Property and Genetic Resources, Traditional Knowledge and Folklore. https://www.wipo.int/tk/en/igc/index.html

WIPO (2018b) Genetic resources. WIPO Intergovernmental Committee on Intellectual Property and Genetic Resources, Traditional Knowledge and Folklore. https://www.wipo.int/tk/en/ genetic/

Wolt JD, Wang K, Yang B (2016) The regulatory status of genome- edited crops. Plant Biotechnol J 14(2):510–518; PMID: 26251102. https://doi.org/10.1111/pbi.12444

WTO (2018) International statement on agricultural applications of precision biotechnol- ogy. Communication from Argentina, Australia, Brazil, Canada, the Dominican Republic, Guatemala, Honduras, Paraguay, the United States of America and Uruguay

Zetsche B, Gootenberg JS, Abudayyeh OO, Slaymaker IM, Makarova KS, Essletzbichler P, Volz SE, Joung J, van der Oost J, Regev A, Koonin EV, Zhang F (2015) Cpf1 is a single RNA-guided endonuclease of a class 2 CRISPR-Cas system. Cell 163(3):759–771. https://doi.org/10.1016/j. cell.2015.09.038

Chapter 3
A Comprehensive Solution for Agriculture 4.0

Abstract This chapter contains a complete explanation of the system that is proposed as a solution. It begins with a description of the problem and argues that it was due to the application of models that responded to the noncooperative, competitive, and mainly zero-sum game theory, without any consideration for the existing relationships between all players. A solution based on a convex superadditive cooperative model of relationships among three sectors or players is proposed, and the balance and transfer of profits among them is presented in the form of a table. Mathematical formulas are provided to sustain the model.

Keywords Game theory · Cooperative game theory · Intellectual property rights · Biotechnological inventions · Plant genetic resources · Balance and transfer · Participant utilities

3.1 The Background of the Problem

PBRs, patents, PGRs, and biosafety are included in a set of separate international, regional, and national legal documents that are actually linked.

There are theoretical studies, ideas, projects, and even pieces of legislation that have tried to find bridges and ways of supplementing and mutual support, at least between IPRs and treaties on PGRs regulating access to any biological material containing functional units of the inheritance, but which have had no impact, or whose negative or positive impact is unknown (Correa 2009).

The tension between IPRs and PGRs is old: IPRs protect any creations of the human mind, and PGRs protect what is in nature. In connection with the principle of the law of nature, theory holds that in principle no IPRs would be valid to be applied on a genetic resource, even if it is known that there are certain cases of legislations that have permitted the patenting of certain PGRs restricting their use by third parties (Correa 2009). But theory holds that, conversely, it is possible to apply breeder rights and patents on plant varieties and biotechnological inventions developed via the access to and use of PGRs. If we add the possibility of adding transgenes to this possibility exceeding sexual reproduction barriers, both breeder rights/patents interface points and biosafety points emerge as an immediate consequence.

M. A. Rapela, *Fostering Innovation for Agriculture 4.0*,
https://doi.org/10.1007/978-3-030-32493-3_3

53

Very early on, it had been stated that in connection with PGRs for plant breeding purposes, a crystal-clear distinction between physical access to them and intangible property was required (Correa 1995). Also long ago, Leskien and Flitner (1997) and more recently Correa et al. (2015) have provided a general outlook and discussed potential supplementing elements for an interface. One example was the recognition of farmers' rights, which, if included in a system for the protection of plant varieties, may help to conciliate the different types of breeders of plant varieties. These authors have considered the possibility of extending the traditional UPOV protection scope to other nonuniform Heterogeneous Plant Varieties (HPVs). They also supported the idea that UPOV-type conventions are not the only ones to be considered, and that there is a wide range of potential systems consistent with TRIPS. We agree with this, but unfortunately there has been no special legislation proposed to support these ideas.

Given the inconsistencies of the CBD, the approval of the ITPGRFA opened up a window with more possibilities to find what in principle was in better consistency with IPRs. The ITPGRFA, applied exclusively on any genetic resources aimed at food and agriculture, is more specific than the CBD and could be thought of as a platform over which the detailed structure of a PGR international policy consistent with IPRs could be built. Even so, the structure would depend on a variety of political, economic, and scientific influences that would likely make it impracticable (Sullivan 2004). Along that line, Oguamanam (2006) has held that the ITPGRFA is inadequate to moderate any conflicts resulting from intellectual property regimes in PGRs. This is so because neither agrobiotechnology nor traditional agricultural knowledge can ensure by themselves that culturally acceptable food be accessible all over the world. This author suggests a dual approach of inequalities in the current intellectual property system. First, developing countries can use domestic legislation to globally define farmers' rights beyond ITPGRFA and, second, the body enforcing the treaty can prioritize the interpretation of the treaty as regards the expectations of developing countries. Onwuekwe (2004) said that the classification of PGRs as common heritage of mankind has not stopped sparking disputes in developing countries that have such resources and industrialized countries with advanced biotechnology using them to incorporate them into new products. The evolution or outcome of multilateral business negotiations, he says, is beneficial because it has given developing countries the opportunity to negotiate the future status of traditional knowledge in connection with IPRs, but there is no guarantee that the outcome is not worse than the initial situation.

A clear example of this can be found when analyzing the work of the WIPO Intergovernmental Committee on Intellectual Property and Genetic Resources, Traditional Knowledge and Folklore. This committee is leading negotiations with the purpose of reaching an agreement about one or several international legal instruments ensuring the effective protection of traditional knowledge, traditional cultural expressions, and genetic resources (WIPO 2018a).

The sessions of this Committee include roundtables of native representatives and presentations by experts who represent those native and local communities, and the purpose is to supplement the frameworks to access and participate in the benefits established by the CBD, the NP, and the ITPGRFA. In connection with IPRs, the Committee is particularly analyzing the following: (a) prevention of incorrect granting of patents, based on the fact that patents on inventions based on PGRs or developed based on PGRs that do not meet the current requirements of novelty and inventive activity must be restricted, (b) ensuring and following up on frameworks for access to and participation in benefits (WIPO 2018b).

By April 2018, the Committee had made progress in drafting a consolidated document in connection with IPRs and PGRs, but its content is a long concatenation of texts in square brackets over which no consensus could be reached (WIPO 2018c). In connection with these challenges for the Committee, Oguamanan (Oguamanam 2018) has stated that the challenges are an opportunity to reexamine the imperative of public domain with the purpose of advocating an inclusive and multicultural jurisprudence.

When attempts were made to apply economic theory to demonstrate that assigning IPRs is important based on efficiency and equity reasons to the matter of conservation of PGRs and in the terms of participation in the benefits, those attempts resulted in failure. As stated by Swanson and Göschla (2000), as there are transaction costs in an industry, the location of an assignment of property rights is a key factor determining the incentives for efficient investment levels. In the context of PGRs, this does not mean that IPRs in the seed industry may not have sufficient effects to generate incentives to provide appropriate amounts of PGRs to the research and development sectors, which are the base of these industries. It has also been stated that an important step for supplementation would be establishing a requirement to disclose the source of any genetic resource or traditional knowledge in applications for property rights with the purpose of adequately protecting the rights of native and local communities (Oberthur et al. 2011), which has not been attained.

In some cases, the supplementation attempts have gone beyond the mere theoretical development. For example, the African Model Law for the Protection of the Rights of Local Communities, Farmers and Breeders, and for the Regulation of Access to Biological Resources of 1998 (OAU 1998), and the Protection of Plant Varieties and Farmers' Rights Act of India of 2001 (India 2001) expressly protect both the right of the breeder and that of the farmer (which results from the ITPGRFA), and therefore includes not only a protection of varieties created by man, but also varieties that result from the empirical selection made by farmers. These laws include provisions allowing communities to claim for compensations, facilitate the registration of native varieties that do not meet the four UPOV requirements (novelty, distinctness, uniformity, and stability), and include provisions on patent law, such as the obligation to disclose the source and the geographical origin of the base biological material (Correa et al. 2015). Given the rigid structure of international treaties, as could be expected, both projects have been rejected by the assemblies of the World Intellectual Property Organization and the UPOV, preventing India from entering the UPOV in spite of its express request to be part of it. The reason alleged is that the previous knowledge by communities and the potential products where they are located must be protected by special laws, and not by UPOV-type laws protecting new and distinct intellectual creations derived from the application of science and technology (Rapela 2016).

In addition to the rejection mentioned, both the African Model Law and the Indian law were drafted bearing in mind social, technical, and productive contexts which are dramatically different from countries such as Argentina, Brazil, Mexico, United States, Canada, or most of the European Union (Rapela 2016). Both include debatable approaches to the resolution of the complementarity and tension between IPRs and PGRs, and are expressly anti-patents, which leaves a zone of full doubt regarding the protection of biotechnological inventions (Dutfield 2008).

Partly in the same direction as the laws mentioned, the People's Republic of China (which has the second largest domestic seed market in the world) renewed its legislation on seeds and breeder rights as from January 1, 2016 (Liu and Zhang 2016). The new text virtually left legislation at the level of the UPOV Convention of 1978, but it only integrated two new aspects—PGRs and biosafety on transgenic plants. In connection with the first, the new legal body transfers the responsibility for PGRs to provinces and regions under a procedure that is highly controlled and regulated by the Government. As regards biosafety, the law has established that the selection, experimenting, validation, and popularization of transgenic plant varieties will be submitted to previous safety assessment and strict monitoring, import, export, and labeling measures will be adopted.

Neither the African Model Law nor the Indian or Chinese laws treated the patent/breeder rights interface or considered that this interface should be taken into account as an element to be regulated, adding even more questions about the practical usefulness of these regulations for the situation of agro exporting countries and based on the newest technological developments. In turn, in the case of the Indian law, recent studies reveal the concern of the private sector about the effects that have already been seen in the long term, including, because of its unprecedented nature, the issue of false claims of property rights because of the lack of reliable databases on the existing varieties (Venkatesh et al. 2016). None of these legal texts addressed, not even implicitly, the problem resulting from products developed via NBTs.

It is necessary to highlight that we do not consider that it is incorrect to have as an objective to claim and incorporate farmers' rights on the varieties developed by them, or to extend the coverage of IPRs to varieties that are not uniform and stable, or the provisions allowing communities to claim for compensations on PGRs that they have kept for dozens, and sometimes hundreds or thousands, of years. Quite the opposite: that approach is correct.

What is incorrect is to try to exacerbate tradition against innovation, to attempt to tie the future with past laws and not to apply a comprehensive vision of the problem.

For example, the genes of toxins Bt present in soil microorganisms incorporated via transgenesis to corn, soybean, cotton, and many vegetable varieties, are part of the genetic inheritance that went through generations and has reached us. This is tradition.

But these genes could not have been incorporated into modern varieties had it not been because it was possible to identify them, isolate them, clone them in functional genetic constructions, insert them in uniform and stable plant varieties and attain their appropriate level of expression, all with the tools of plant breeding and modern biotechnology. Such progress meets the requirements for breeder rights and patentability in almost all countries in the world. This is innovation.

Tradition and innovation must be balanced, and this is what cannot be seen in the legislative texts mentioned.

Separate from the multiple IPRs regimes in force in the world, there is a universe of PGRs that is not being used. New tension emerges in this context, and traditional knowledge of local and native communities—including agriculture, stockbreeding, medicine, biological resources, etc.—is facing increasing challenges in any international set of laws. Their access and use makes them more vulnerable, especially in connection with misappropriation and misuse. Several countries[1] have tried to protect such traditional knowledge and there have also been regional efforts in this sense Ouma 2017).[2] The problem is that these national and regional laws protect this knowledge without limits, and they are only effective in countries that have passed them (Ouma 2017).

Finally, evidence shows that considering only the effects of the TRIPS and its resulting laws on the productivity of crops, the results are confusing. For example, based on a new index to measure the strength of IPRs on the productivity of several varieties and in several countries, Campi and Nuvolari (2015) and Campi (2016) found positive and negative effects in all directions. The conclusion of these authors is that a single intellectual property system could not be the solution and it is necessary to find answers to each model of productive development.

We do not agree with that position here. The application of differentiated strategies for each productive model could result in a much more serious situation than the current one, as there would be as many laws as countries. The same can be extended to the protection of traditional knowledge, including PGRs.

Other experts have identified the need to establish a consistent international regime for their preservation via a binding international treaty that is above national and even regional efforts (Ouma 2017). The proposal of this book goes in that direction and is therefore opposite to the implementation of differentiated strategies. The proposal presented is based on the development of a comprehensive and harmonious system covering in an inclusive manner the aspects of breeder rights, patents, PGRs, also considering the interests of all parties and sectors. To sum up, this is a comprehensive proposal and it is also explicit to be analyzed, corrected, and potentially applied in practice.

---

[1] Costa Rica, Kenya, Peru, Zambia.

[2] For example, the Swakopmund Protocol on the Protection of Traditional Knowledge and Expressions of Folklore adopted in August 2010 by 19 members of the African Regional Industrial Property Organization (ARIPO).

3.2   Game Theory

Resorting to game theory, let us now analyze the same situation from another perspective.

In real life, people and groups of people may be in situations in which their interests are partially or fully inconsistent. This may remain so, or the conflict may be somehow solved. Game theory is the formal study of decision-making where several players (persons, groups of persons, industries, institutions, companies, or any combination of them) must take decisions potentially affecting the interests of other players and, therefore, the "game" is the description or formal model of a strategic situation.

Game theory models have been developed to most adequately represent the reality applying to the analysis the model that best fits the behaviors and decisions that participants actually make, given the incentives and restrictions they face. Models may be classified into a high degree of detail, but they generally include two categories: (a) in noncooperative game theory, players (often only two of them) make their decisions independently and with the purpose of maximizing their own interests without considering others' interests, they have no mutual commitments, there is no gain transfer, and they cannot make any previous arrangements; (b) conversely, in cooperative game theory, players have the possibility of communicating and two or more players do not compete—instead, they collaborate establishing arrangements allowing for the formation of coalitions with the same objective in mind and, therefore, they all win or lose, detailing what utility a potential group or coalition may obtain by virtue of cooperation of their members. Cooperative games include situations in which utilities may be transferred in any manner among coalitions, and situations in which utility cannot be transferred. Therefore, cooperative games may be TU or transferable-utility games, or NTU, that is, nontransferable utility games. In any event, what the theory does not make explicit is the procedure whereby the coalition is formed (Turocy and von Stengel 2001; Peleg and Sudhölter 2007).

Game theory has had a very important role in negotiation processes—it has been applied in an infinite number of areas, including the best distribution of PRG resources. Chantreuil and Cooper (2005) say that, in the case of PGRs, negotiation solutions such as that of John Nash or other noncooperative solutions are not adequate tools for models with more than two players where any number of free decisions may be taken and where manipulation of coalitions may also occur.

In this regard, Aoki (2009) states that the problem is complex because in the application of game theory to PGRs the particular interests of sectors on common shared resources are involved. This author highlights the conclusions of Ostrom (1990), who mentions three elements (to which a fourth may be added) to be considered and which we will fully apply in this case:

(a) The "tragedy of the commons" (Harding 1968), which occurs when rational and well-intentioned individuals, acting based on their own interests (let us assume, for this case, local communities, seed companies, and biotechnology consortiums), ultimately end up destroying a shared and limited resource (PGR), even

if faced with evidence that such course of action does not benefit anybody in the
long term

(b) The "prisoner's dilemma," a key problem of game theory, which permits to
explain the behaviors of competing players (e.g., germplasm developing companies and biotechnological consortiums) in oligopolistic markets, in which
companies do not cooperate among themselves and maintain their short-term
individual interests even if this action goes against their own interests in the
long term

(c) The "logic of collective action" (Olson 1971), where only individual and selec-
tive incentives drive rational players (local communities, seed companies, and
biotechnological consortiums) to act based on group spirit; in other words, individuals act collectively just to provide private goods, but not to provide any
public goods

(d) The "rise and decline of nations" (Olson 1982), whereby the policies resulting
from individual actions by different players only result in "pressure groups,"
whose resulting strategies tend to be protectionist and contrary to technological
innovation

In the context of PGRs, but which we extend to the entire situation, Chantreuil
and Cooper (2005) state that given the large number of potential players in these
agreements (the authors assume that the countries are the players), as well as the
general desire to adopt coalition solutions to attain such agreements, the cooperative
game theory emerges as a technique especially fit for those agreements, as opposed
to noncooperative game strategy. They have developed a modeling of cooperative
game theory for n-players with a number application to determine the fair and equitable participation in benefits. Ostrom (1990) recognizes that applying IPRs is not
easy or simple in connection with common resources such as PGRs, and the case is
more difficult for plant germplasm. But to the extent that no incentives are generated to create or maintain these collective resources, the situation will be increasingly worse.

Analyzing the African Law and the laws of India and China, one can see that they
are consistent with a modeling adjusted to the noncooperative game theory, in which
the players defined a priori came down to only two: local and ethnic communities
with PGRs and seed companies. Consortiums developing biotechnological inventions, whatever their form, were left outside the analysis (game). Moreover, in this
kind of bilateral game, the strategy applied, in addition to being competitive, was
basically zero-sum (games which are strictly competitive and with opposing interests between the parties, in which the gain or loss of one entails the gain or loss of
another participant), and not one of non-zero-sum games (where the interests of the
parties are not fully opposing and the situation might be win-win for both
participants).

In any event, the new legislation triggered a critical shift in the international
paradigm of PGRs, changing from "common heritage of mankind" to "State's sovereignty." Within the framework of game theory, there was a shift from value $v = 0$

for the resource, to a positive-sign value or, in other words, the PGR came to have a v, and the transfer of utilities became a factor to be considered.

The consequences of all this were not minor. The paradigm shift resulted in that PGRs could have value (which is not questioned), but this probably led to an excessive value, as the PGR is unknown beforehand. One does not know whether the PGR exists or not, it does not materialize until isolated, introduced, and developed in a plant variety, and this can only be done by a professional team, whether public (universities or research centers) or private (seed companies). These particularities seemed to be circumstantial and even secondary. The prevailing approach was a set of criteria resulting from competitive strategies such as focusing on one's own gain, winning at all costs and as much as possible (zero-sum games), hiding information, system bureaucratization, own interest over the interest of others, exacerbation of bargaining power, and emotional aspects, indifference to negative or positive externalities, total disregard for the long-term impact.

Value v, "benefit," became a key issue, and it is worth citing the extensive annex to the NP to exemplify this:

(1) Monetary benefits may include, but not be limited to: (a) Access fees/fee per sample collected or otherwise acquired; (b) Up-front payments; (c) Milestone payments; (d) Payment of royalties; (e) Licence fees in case of commercialization; (f) Special fees to be paid to trust funds supporting conservation and sustainable use of biodiversity; (g) Salaries and preferential terms where mutually agreed; (h) Research funding; (i) Joint ventures; (j) Joint ownership of relevant intellectual property rights. 2. Non-monetary benefits may include, but not be limited to: (a) Sharing of research and development results; (b) Collaboration, cooperation and contribution in scientific research and development programmes, particularly biotechnological research activities, where possible in the Party providing genetic resources; (c) Participation in product development; (d) Collaboration, cooperation and contribution in education and training; (e) Admittance to ex situ facilities of genetic resources and to databases; (f) Transfer to the provider of the genetic resources of knowledge and technology under fair and most favourable terms, including on concessional and preferential terms where agreed, in particular, knowledge and technology that make use of genetic resources, including biotechnology, or that are relevant to the conservation and sustainable utilization of biological diversity; (g) Strengthening capacities for technology transfer; (h) Institutional capacity-building; (i) Human and material resources to strengthen the capacities for the administration and enforcement of access regulations; (j) Training related to genetic resources with the full participation of countries providing genetic resources, and where possible, in such countries; (k) Access to scientific information relevant to conservation and sustainable use of biological diversity, including biological inventories and taxonomic studies; (l) Contributions to the local economy; (m) Research directed towards priority needs, such as health and food security, taking into account domestic uses of genetic resources in the Party providing genetic resources; (n) Institutional and professional relationships that can arise from an access and benefit-sharing agreement and subsequent collaborative activities; (o) Food and livelihood security benefits; (p) Social recognition; (q) Joint ownership of relevant intellectual property rights.

Conversely, the ITPGRFA had a significantly pragmatic approach from scratch, as it explicitly defined that the facilitated access to a PGR is by itself the key benefit, and it implicitly entailed that incorporating a PGR to a product while at the same time leaving that product free for future research and genetic breeding is the compensatory benefit.

In discussing sharing of benefits and other commercialization of PGRs, Article 13.2(d) of the ITPGRFA, it was established that a recipient who commercializes a product that incorporates material accessed from the Multilateral System, "shall pay . . . an equitable share of the benefits arising from the commercialization of that product, *except* whenever such a product is available without restriction to others for further research and breeding" (emphasis added). The ITPGRFA implicitly recognized the "open source" nature of the UPOV Convention (one of its key pillars) and introduced this addition.

However, this benefit-sharing under the ITPGRFA establishing that $v = 0$ only if the PGR that was accessed is within an open-source, UPOV-type model, eventually created the greatest challenge for the Treaty—virtually nobody paid, and the multilateral ITPGRFA system became underfunded.

The three conclusions by Ostrom (1990), and the fourth added before, were verified in all of these events. While noncooperative strategies try to maximize utility and do not seek adequate solutions, but the prediction of results based on foreseeable behaviors, in these cases those strategies resulted in inefficient balances.

In contrast, a cooperative or coalitional game may have different characteristics and reach efficient balance, provided that it is possible to understand the way of modifying the functions for transferring participants' utilities.

Following classical theories of cooperative game theory (Magaña Nieto 1996; Peleg and Sudhölter 2007; Puga 2013), first a definition should be made regarding whether or not the utility is transferable, assuming in our case that it is given that the potential utility may be shared in any manner among its members.

A cooperative game with transferable utility is a pair *(N, v)*, containing, to start with, a finite number but higher than two players:

$$N = \{1,2,\ldots,n\} \tag{3.1}$$

Every element of set N is a player and every subset of N is a coalition, also having a characteristic function associating every subset S *of J* (or coalition) with a real number $v(S)$, which is the value of the coalition. If there is no coalition, that is, players do not establish any kind of cooperation among them, the value generated is zero:

$$v(\varnothing) = 0 \tag{3.2}$$

Therefore, game $G = (N, v)$ is a cooperative or coalitional game with transferable utilities to the extent that N and v are well defined.

In our case, we have defined $N = \{1, 2, 3\}$

- $N1$ = states, provinces, local, ethnic, or native communities with PGRs.
- $N2$ = seed companies and/or public or mixed entities involved in the creation and development of plant varieties.
- $N3$ = consortiums developing biotechnological inventions. These may be academic centers, private, or mixed entities.

The function uniting all the possibilities is the following:

$$v = \{v(1), v(2), v(3), v(12), v(13), v(23), v(123)\} \tag{3.3}$$

If a cooperative, monotonous-type game *(N, v)* is established, the benefit or gain from the coalition is not reduced if the number of players increases:

$$v(S) \le v(T) \tag{3.4}$$

The cooperative game *(N, v)* will be additive if:

$$v(S \cup T) = v(S) + v(T), \text{for every } S, T \in 2N, S \cap T = \varnothing \tag{3.5}$$

The cooperative game will be superadditive if:

$$v(S \cup T) \ge v(S) + v(T), \text{for every } S, T \in 2N, S \cap T = \varnothing \tag{3.6}$$

Both in the additive and in the superadditive game, the benefit of the union of two coalitions is equal to or higher than the addition of individual benefits, but, in the superadditive game, players have incentives to form the "great coalition," that is, the coalition containing *N* players as there is no other form of coalition or grouping of players attaining more than what the coalition or grouping of all may attain, verifying the following:

$$\begin{aligned} v(\{1\} + v(\{2\} \le v(\{1,2\}), v(\{1\}) + v(\{3\}) \le v(\{1,3\}) \\ v(\{2\} + v(\{3\} \le v(\{2,3\}), v(\{1\}) + v(\{2,3\}) \le v(\{1,2,3\}) \\ v(\{2\} + v(\{1,3\} \le v(\{1,2,3\}), v(\{3\}) + v(\{2,3\}) + v(\{1,2\}), \le v(\{1,2,3\}) \end{aligned} \tag{3.7}$$

The superadditivity condition is not the only one necessary for the optimal model.

Let us assume that the 1,2 coalition in our example offers a benefit equal to coalition 1,3, but that coalition 2,3 (i.e., among developers of plant varieties and biotechnologists) gives more benefits. In this case, the function of this game *(N, v)* with *N* = {1, 2, 3} is defined by:

$$\begin{aligned} v(1) = 0, v(2) = 0, v(3) = 0, \\ v(1,2) = 1, v(1,3) = 1, v(2,3) = 6 \\ v(1,2,3) = 6 \end{aligned} \tag{3.8}$$

This is a superadditive game, but it is not convex as:

$$v(1,2,3) + v(2) \ge v(2,3) + v(1,2) \tag{3.9}$$

The optimal game may be in a superadditive cooperative game strategy, which also meets the convex condition requirement, that is, when the contribution of a player to the coalition does not decrease when more players are incorporated. The function of this game *(N, v)* with $N = \{1, 2, 3\}$ is defined by:

$$v(1) = 0, v(2) = 0, v(3) = 0,$$
$$v(1,2) = 2, v(1,3) = 2, v(2,3) = 2 \qquad (3.10)$$
$$v(1,2,3) = 9$$

Verifying that:

$$v(1), v(2), v(3) \le v(1,2), v(1,3), v(2,3) \le v(1,2,3) \qquad (3.11)$$

Every convex game is superadditive, but the opposite is not necessarily true. The conclusion reached points to the need of creating a convex superadditive cooperative model of relationship among the three sectors or players, but that requires devising a new balance and utility-share system.

The basic points of the model are the following:

(a) Every player has an equivalent "importance." This does not mean that the potential distribution of cooperation benefits should be equivalent.[3]
(b) Every coalition of two players (1, 2), (1, 3), (2, 3) has an equivalent importance.
(c) Every player has to give in something in exchange for something else.
(d) The joint coalition containing *N* players must be the most beneficial for everybody.
(e) The success or failure of the model must be measurable via an externality, which in this case is the annual genetic progress rate for the main crops.

The model is not intended to be a "salvation board" for products derived from new technologies, or any products derived from the work of local, native, or ethnic communities that are now virtually outside the scope of protection, but to incorporate them into an integrated system.

A germplasm integrated system must contemplate all players and all these aspects and possibilities. In turn, they must stimulate innovation and share knowledge.

---

[3] Developing a single transgenic event and taking it to the condition of worldwide commercialization is virtually the same as the technical budget of all PGR centers in the world.

Table 3.1 System of balance and transfer of participants utilities

Player	Gives	Receives	Expected positive externality
N1 = states, provinces, local, ethnic, or native communities with PGRs.	They have to register their products and make them available via a simple, fast, and free process. The definition of genetic resource based on tangible components is abandoned.	The rights acquired over products derived from the application of knowledge, traditional practices, and innovation of native peoples, local and ethnic communities are affirmed. Nothing that is preexisting in nature may be subject to a property right. No protected germplasm may be natural germplasm. Any protected germplasm becomes natural germplasm after the protection period finishes. Both plant varieties or plant traditional developments may access ownership titles. Germplasm becomes tangible structure and intangible information.	Simple and fast access and use of genetic resources. All natural germplasm is made available for breeding, and all protected germplasm becomes natural germplasm after a reasonable period.

Table 3.1 (continued)

Player	Gives	Receives	Expected positive externality
$N2$ = seed companies and/or public or mixed entities involved in the creation and development of plant varieties.	The current definition of "variety," qualified as such due to its phenotype, is abandoned. The period to protect plant varieties is shortened by several years.	The plant variety is defined by its genotype and its phenotype. For the purposes of being qualified as a new variety, the extent of the phenotypical change is not important. The phenotypical change may be minimal or extensive, provided that it leads to a qualitative character change that is expressed. Heterogeneous plant varieties (HPVs) are defined and become known with an ownership title. It is not necessary for them to observe the uniformity requirement. Heterogeneous plant varieties that nowadays are not covered by any protection system are added to the list of protectable creations.	The new definition of variety is aimed at the creation of more variability. The protection of HPVs balances the scope of rights over other important plant breeding developments. The limitation of protection leaves adequate room for breeders to recover their investments and is an incentive for fast renewal of varieties or HPVs.
$N3$ = consortiums developing biotechnological inventions. These may be academic centers, private, or mixed entities.			

(continued)

Table 3.1 (continued)

Player	Gives	Receives	Expected positive externality
Coalition ($N1, N2$)	$N1$: They fully facilitate access to and use of genetic resources and traditional knowledge. $N2$: They make a prior payment, regardless of whether the resource or knowledge has been used. Implicit recognition is given to all the breeding history before inheritance and genetic knowledge.	$N1$: Local communities and source centers receive previous compensation, regardless of the use of the resource or knowledge. $N2$: They access genetic resources and traditional knowledge without restriction and in an easier manner.	All germplasm is made available for research and development. Crop innovation and productivity rate increased.
Coalition ($N1, N3$)	$N1$: They fully facilitate access to and use of genetic resources and traditional knowledge. $N3$: They make their R&D capabilities available to use knowledge and resources in products, giving them value. The payment for the resource is made via $N2$'s contribution.	$N1$: Local communities and source centers receive previous compensation, regardless of the use of the resource or knowledge. $N3$: They access genetic resources and traditional knowledge without restriction and in an easier manner.	All genomic resources are made available with possibilities of becoming biotechnological inventions. Crop innovation and productivity rate increased.
Coalition ($N2, N3$)	They leave aside all restrictions on the research and experimenting phase.	Access to all germplasm.	Innovation rate increased.

(continued)

Table 3.1 (continued)

Player	Gives	Receives	Expected positive externality
Large Coalition (N1, N2, N3)	Adhering to separate conventions or treaties is abandoned. A "narrow" definition of germplasm is abandoned. Patents, breeder rights, plant genetic resources, and biosafety no longer depend on offices, and the following table of balance and transfer of participants' utilities is proposed: separate ministries.	A single comprehensive treaty to cover the access, use, and shared benefit of PGRs, protection of plant varieties, protection of biotechnological inventions, and biosafety. "Germplasm" covers any kind of plant organism and inheritance functional units. Patents, breeder rights, plant genetic resources, and biosafety are the issues in the treaty and are coordinated from a single office, the "Governing Body."	Recovery of productivity rates in all crops. Food biodiversity increased. Incentive for genetic improvement in all crops.

The empirical plant breeding applied by local communities, scientific plant breeding, whether conventional or assisted by molecular markers and any transgenic or non-transgenic modern biotechnology, are the tools to be applied.

3.3   Balance and Transfer Table

Based on the cooperative game model, we propose in Table 3.1 a system of balance and transfer of participants' utilities.

References

Aoki K (2009) Free seeds, not free beer: participatory plant breeding, OpenSource seeds, and acknowledging user innovation in agriculture. Fordham Law Rev 77(5):Article 9. Available at: http://ir.lawnet.fordham.edu/flr/vol77/iss5/9

Campi M (2016) The effect of intellectual property rights on agricultural productivity. Agric Econ 48:1–13

Campi M, Nuvolari A (2015) Intellectual property protection in plant varieties. A worldwide index (1961-2011). Res Policy 44(4):951–964

Chantreuil F, Cooper JC (2005) Modeling the impacts of Bargaining Power in the Multilateral Distribution of Agricultural Biodiversity Conservation Funds. In: Cooper JC. Lipper LM, Zilberman D (eds) Agricultural Biodiversity and Biotechnology in Economic Development. Natural Resource Management and Policy, vol 27. Springer, Boston, MA.

Correa CM (1995) Sovereign and property rights over plant genetic resources. Agric Hum Values 12(4):58–79

Correa CM (2009) Trends in Intellectual Property Rights relating to Genetic Resources for Food and Agriculture. Background Study Paper No 49 October 2009. Commission on Genetic Resources for Food and Agriculture

Correa CM, Shashikant S, Meienberg F (2015) Plant variety protection in developing countries: a tool for designing a sui generis plant variety protection system: an alternative to UPOV 1991. Association for Plant Breeding for the Benefit of Society (APBREBES) and its member organizations: Berne Declaration, Development Fund, SEARICE, Third World Network. Available at http://www.apbrebes.org/news/new-publication-plant-variety-protection-developing-countries-tool-designing-sui-generis-plant. Accessed 24 Oct 2017

Dutfield GM (2008) Turning plant varieties into intellectual property: the UPOV convention. In: Tansey G, Rajotte T (eds) The future control of foods: a guide to international negotiations and rules on intellectual property, biodiversity and food security. Earthscan, London / Sterling, VA, IDRC. ISBN: 978-1-84407-429-7, pp 27–47

Harding G (1968) The tragedy of the commons. Science 162(3859):1243–1248

India (2001) The protection of Plant Varieties and Farmers' Rights. http://lawmin.nic.in/ld/P-ACT/2001/The%20Protection%20of%20Plant%20Varieties%20and%20Farmers%E2%80%99%20Rights%20Act,%202001.pdf

Leskien D, Flitner M (1997) Intellectual property rights and plant genetic resources: options for a sui generis system. Issues in genetic resources no. 6, June 1997. International Plant Genetic Resources Institute, Rome

Liu C, Zhang W (2016) Recent advances in protection of new varieties of plants. IP News and Cases, Liu Shen and Associates, http://www.liu-shen.com/Content-2415.html

Magaña Nieto A (1996) Formación de coaliciones en los juegos cooperativos y juegos con múltiples alternativas. Departament de Matematica Aplicada II. Universitat Politecnica de Cataunya

OAU (1998) Organization of African Unity (OAU) African Model Legislation for the Protection of the Rights of Local Communities, Farmers and Breeders, and for the Regulation of Access to Biological Resources. http://www.wipo.int/wipolex/en/text.jsp?file_id=252153

Oberthur S, Gerstetter C, Lucha C, McGlade K, Pozarowska J, Rabitz F, Tedsen E (2011) Intellectual Property Rights on Genetic Resources and the fight against poverty. European Parliament, Directorate General for External Policies. Policy Department. European Parliament, 2011

Oguamanam C (2006) Intellectual property rights in plant genetic resources: farmers' rights and food security of indigenous and local communities. Drake J Agri Law 11:273–305

Oguamanam C (2018) Wandering footloose – traditional knowledge and the "public domain" revisited. J World Intellect Prop 21(5–6):306–325

Olson M (1971) The logic of collective action: public goods and the theory of groups, Harvard University Press, Cambridge, Massachusetts, 1st ed 1965, 2nd ed 1971

Olson M (1982) The rise and decline of nations: economic growth, stagflation, and social rigidities. Yale University Press, New Haven

Onwuekwe CB (2004) The commons concept and intellectual property rights regime: whither plant genetic resources and traditional knowledge? Pierce Law Rev 2:65. Available at http://scholars.unh.edu/unh_lr/vol2/iss1/7

Ostrom E (1990) Governing the commons: the evolution of institutions for collective action. Cambridge University Press, Cambridge, UK

Ouma M (2017) Traditional knowledge: the challenges facing international lawmakers. WIPO Magazine, February 2017. This article is based on the keynote address by Dr. Ouma at the WIPO Seminar on Intellectual Property and Traditional Knowledge in Geneva, Switzerland, in November 2016

Peleg B, Sudhölter P (2007) Introduction to the theory of cooperative games. Theory and decision library, series C: game theory, mathematical programming and operations research, 2nd edn. Springer, Berlin

Puga MS (2013). Estructuras jerárquicas y juegos cooperativos con utilidad transferible. Universidad de Santiago de Compostela, Universidade Da Coruña, Universidad de Vigo. 11 de enero 2013. Directoras: Fiestras Janeiro MG, Sánchez Rodríguez, E

Rapela MA (2016) Ley 20.247 de Semillas y Creaciones Fitogenéticas: las razones para su actualización y los proyectos bajo análisis en Argentina. Revista Interdisciplinaria de Estudios Agrarios, Facultad de Ciencias Económicas, Universidad de Buenos Aires 45:69–98

Sullivan SN (2004) Plant genetic resources and the law: past, present, and future. Plant Physiol 135(1):10–15. https://doi.org/10.1104/pp.104.042572

Swanson T, Göschla T (2000) Property rights issues involving plant genetic resources: implications of ownership for economic efficiency. Ecol Econ 32(1):75–92

Turocy TL, von Stengel B (2001) Games theory. London School of Economics CDAM Research Report LSE-CDAM-2001

Venkatesh P, Sekar I, Jha GK, Singh P, Sangeetha V, Pal S (2016) How do the stakeholders perceive plant variety protection in Indian seed sector? Curr Sci 110(12):25

WIPO (2018a) WIPO Intergovernmental Committee on Intellectual Property and Genetic Resources, Traditional Knowledge and Folklore. https://www.wipo.int/tk/en/igc/index.html

WIPO (2018b) Genetic resources. WIPO Intergovernmental Committee on Intellectual Property and Genetic Resources, Traditional Knowledge and Folklore. https://www.wipo.int/tk/en/genetic/

WIPO (2018c) Consolidated document relating to intellectual property and genetic resources. WIPO/GRTKF/IC/36/4. April 10, 2018

Chapter 4
Plant Germplasm Integrated System

Abstract This chapter contains the complete development of the Plant Germplasm Integrated System, based on a binding treaty with 21 articles. Almost all articles have a footnote call in order to expand or argue each point. In many cases, specific mention is made that the point is a "governing principle" of the "System." When a "governing principle" is mentioned, it responds to the Table of Balance and Transfer, from Chap. 3, and to a cooperative game.

Keywords Plant germplasm · Integrated system · Binding treaty · Intellectual protection · Traditional developments · Plant varieties · Heterogeneous plant varieties · Biotechnological inventions · Microorganisms · Biosafety

Based on the arguments presented in the earlier chapters, we will now present the minimum elements of a "Plant Germplasm Integrated System" (hereafter, the "System") as a comprehensive and inclusive proposal for all sectors related to the protection of heterogeneous plant varieties (HPVs), plant varieties, microorganisms, biotechnological developments, genetic resources, and biosafety.

Almost all articles have a footnote elaborating every point. In many cases, express mention is made to the effect that the point is a "governing principle" of the System. Every time a governing principle is mentioned, that principle responds to the balance table in Chap. 3 and to a cooperative game.

4.1 Minimum Requirements

4.1.1 Article 1. Purposes

1. With the purpose of increasing the innovation rate in the genetic breeding of plant species, the Plant Germplasm Integrated System (the "System") has five purposes: (a) preserving the existing germplasm and increasing its genetic

variability; (b) promoting the access and sustainable use of germplasm; (c) asserting acquired rights over products derived from the application of knowledge, traditional practices and innovation efforts of native peoples, ethnic and local communities; (d) securing the intellectual property over the new germplasm stimulating agricultural experimentation in connection with plant breeding and modern biotechnology, and (e) establishing the biosafety framework over new germplasm.[1]

2. For such general and specific purposes, it is hereby established that the System is the sole and exclusive way for: (a) the conservation and sustainable use of plant genetic resources and the fair and equitable distribution of benefits derived from their use; (b) obtaining property rights over new germplasm; and (c) the new germplasm's biosafety.[2]

3. Member states undertake to modify their domestic laws in connection with plant genetic resources, breeder rights for the protection of plant varieties, patenting for the protection of biotechnological inventions with plant or microbiological source, and biosafety to make up a legal system that at least agrees with the general and particular purposes of this System.[3]

4. This System will not affect any other obligations of member states that are consistent with it, resulting from the international agreements signed and, especially from the TRIPS Agreement, the Convention on Biological Diversity, and the International Treaty on Plant Genetic Resources for Food and Agriculture, which will be terminated only if they are renounced.[4]

---

[1] The System establishes a single general purpose that responds to the key claim of this book: the maze of regulations applied on plant varieties is affecting research and development in plant genetics and that also affects the crop production increase rate. If this claim is accepted, the intended purpose is to change the current situation restricting this production increase. Five specific purposes result from the general purpose: (a) preservation and conservation of germplasm increasing genetic variability; (b) access to and sustainable use of germplasm; (c) recognition of acquired rights of native peoples and local communities over plant products; (d) stimulating experimentation on plant breeding via the application of IPRs in a wide and comprehensive framework, and (e) establishing the guidelines of a modern biosafety framework. These five purposes are the first *governing principle*.

[2] The System suggests a fully integrated model in a single legal instrument to include regulatory aspects in a broad sense, including any intellectual property, any genetic resources with a plant source, and the biosafety regulatory framework. This is a *governing principle*. Against this backdrop, the treaty is a unique document at the international level.

[3] Adherence to the System entails that every member state will have to pass a national law to make the System effective in their territories. The result of this is that existing laws on breeder rights, patents on biotechnological inventions of plant or microbiological source, plant genetic resources, and biosafety that are not consistent with the System are immediately abrogated. This abrogation is in full in the case of breeder rights for the protection of plant varieties, as the entire scope of the existing legislation is established in this System, and the same applies to biosafety. This is not so for patents and genetic resources, as there are a large number of inventions that fall outside the System and there are non-plant genetic resources.

[4] Based on what has been explained in the above footnote, adherence of a country to the System does not entail that the country renounces the CBD, the TRIPS, or the ITPGRFA.

5. The scope of this System will only have uniform effects within the territory of member states and rights may only be granted, transferred, or terminated in such territory uniformly.[5]
6. For the purposes of applying this System and its administration, a Governing Body is established, whose scope and tasks are detailed under Article 11.[6]

4.1.2 Article 2. Definitions

1. "Germplasm" means any material from the domains of Archaea, Bacteria, and Eukarya (including kingdoms Protista, Fungi, and Plantae), in addition to viruses, which may be used in the genetic breeding of any biological material, including reproductive and plant propagation material containing inheritance functional units and with actual or potential value for food and agriculture, including any living plant organism, both seeds in strict botanic terms, as well as fruits, bulbs, tubercles, shoots, stakes, rootstocks, cut flowers, leaves, stems, any structure made up by plant tissue with the capacity to generate an entire plant in vivo or in vitro, any kind of microorganism or biological material that is natural or has been obtained as a result from modern biotechnology, including cell inclusions, chromosomes, genes, and part of them, as well as the information in digital sequence that all or part of these organisms contain.[7]
2. "Protected germplasm" is any part of germplasm that is protected by any intellectual property right under this System. Any protected germplasm is new by

---

[5] The principle of territoriality of the scope of the System is affirmed. Territory is defined in Articles 4.1 and 4.3. See also footnotes 44 and 46.

[6] Its makeup and tasks are detailed under Article 11.

[7] The concept of germplasm has many meanings in the bibliography, and the most general one is the one applied to designate the genome of any species. The definition of "germplasm" given here is exhaustive and descriptive, with the purpose of establishing that not only every piece of plant matter from components of inheritance functional units, cellular inclusions, all taxonomic groups of microorganisms, including viruses, to complete plants, are within the System, but also the information contained. So, Germplasm becomes a definition of tangible and intangible components. This is a *governing principle*. The expression "information in digital sequence" has been introduced to specify nontangible information, abandoning the definition and accurate scope of the term being discussed in the CBD (CBD 2016). The classification of organisms used here has followed the proposal introduced by Woese and Fox (1977) and Woese et al. (1990), which divides cellular life-forms into archaea, bacteria, and eukaryote domains. Considerations for the eukaryote domain has followed the proposal made by Adl et al. (2005, 2012), which was the conclusion of an international group of experts on the subject matter. In connection with viruses, whether or not they are protected is discussed in international regulations. According to the European Patent Office, the United States system, and the TRIPS, they are protectable matter, and this is the criterion followed in this System.

definition. It is not possible, under any circumstance that "protected germ-plasm" has previously been "natural germplasm."[8]

3. Natural germplasm is that part of the resulting germplasm as a consequence of natural evolution of biological systems and is not covered by any IPRs. Any protected germplasm whose legal protection has terminated or has been declared null for whatever reason is also part of natural germplasm and, there-fore, there cannot be natural germplasm that has previously been protected germplasm.[9]

4. "Plant variety" means any set of plants of a single botanical taxon of the lowest range known of the Plantae kingdom that may be:

   (a) Defined by the genotype and the phenotypical expression of the resulting characters of such genotype or of a certain combination of genotypes
   (b) Distinguished from any other set of plants for having a new genotype or a certain combination of genotypes and for the phenotypical expression of at least a qualitative character of such genotype or combination of genotypes
   (c) Considered as a unit, given its aptitude to propagate without alteration

     A set of plants is formed by full or parts of plants, provided that these parts may generate entire plants.

     The expression of the characteristics mentioned under section (a) must be invariable, considering the logical behavior of living systems.[10]

5. "Heterogeneous Plant Variety (HPV)" means any set of plants of a single botanical taxon of the lowest range known of the Plantae kingdom, which may be:

   (a) Defined by the expression of the resulting characters of a certain combina-tion of genotypes

---

[8] Articles 2.2 and 2.3 differentiate two types of germplasm—protected and natural. This is a *governing principle*. This principle—repeated in other articles—establishes that nothing that is preex-isting in nature may be protected via intellectual property. An interaction between both "germplasms" is established that goes in one sense only. This means that a natural germplasm may have been protected germplasm in one case, but not the other way around. The natural germplasm is made up by the work of nature and cannot be protected via intellectual property.

[9] See footnote No. 22.

[10] The definition of "plant variety" in the treaty departs from UPOV, including the versions of the 1961/1972 Conventions or the 1991 Convention, in establishing that the definition is given by the genotype or combination of genotypes and, jointly, by the expression of the resulting characters of such genotype or combination of genotypes. As there are precedents that two varieties of different genotypes may have the same phenotype, and that two varieties of different phenotypes may have the same genotype (UPOV 2004), the qualification of "different plant variety" will be assigned when genotype and phenotype differ. The extent of the genotypic change (number of modified bases) is not relevant for the purposes of the distinction. This is a key element, as the change in a single nucleotide may make a variety different, provided that this single change produces a differ-ent phenotype via a qualitative change. In an analogous manner, an extensive genotypic change (e.g., in a genotype's silent area) does not make a variety different if such change does not express a phenotypical change. This is a *governing principle* included in the System aimed at promoting the generation of true diversity in the creation of new varieties. The "combination of genotypes" referred to is for the purposes that hybrid varieties fit the definition, which already has the prece-dent of the definition of "variety" under the 1991 UPOV Convention. It is necessary to interpret this definition together with Article 6 of this Treaty about "plant varieties."

(b) Distinguished from any set of plants by the joint phenotypical expression of such genotypes

(c) Considered as a plurality, given its aptitude to propagate without alteration in the distinctive features.[11]

6. Genotype: This is the genetic constitution or composition of an organism determined by the localization and combination of genes in chromosomes.[12]

7. "Phenotype": It is the expression of the genotype of an organism that may be affected by the environment.[13]

8. "Microorganism": Any set of individuals pertaining to the Archaea, Bacteria, Protista, or Fungi kingdoms, in addition to viruses.[14]

9. "Living organisms": Any biological entity capable of transferring or replicating genetic material, including sterile organisms, viruses, and viroids.[15]

10. "Genetically modified living organism": Any living organism with a new combination of genetic material.[16]

11. "New combination of genetic material": This is the stable and simultaneous insertion of one or more genes or DNA sequences being part of a genetic construction that is genetically defined and permanently introduced in the genome of the receiving plant, coming from unrelated species or species that are not compatible with the receiving species. The "null segregant" is not a "new combination of genetic material," that is, the progeny of a GMO in which the only genetic modification has been the insertion of the donor organism's nucleic acid in the genome of the receiving organism, but the nucleic acid of the donor

---

[11] The Treaty defines a new plant grouping other than the "plant variety" that is the "heterogeneous plant variety" (HPV), with the purpose of accommodating genuine technical developments that do not qualify to be "varieties," as they are made up by genotype mixtures, and, therefore, their phenotypical expression will not be uniform. This category includes, for example, local populations and races, geographical compounds, ethnical compounds, etc. This is a *governing principle*. It is necessary to interpret this definition together with Article 7 of this Treaty about "heterogeneous plant varieties."

[12] Usual definition of "genotype."

[13] Usual definition of "phenotype."

[14] Usual definition of "microorganism." Take note of the comments in connection with Article 2.1. In any case and consistent with the spirit and the exhaustive mention, it is not possible for the System to obtain protection via intellectual property in any kind of microorganism. It will be possible to obtain protection over a genetic sequence of them or over a new microorganism created by the human being's action.

[15] Usual definition of "living organism"

[16] Articles 2.10, 2.11, and 2.12 must be interpreted jointly. All of them are a *governing principle*, whose sources come from the definition of NBT products of Argentina's and Israel's biosafety regulatory system and part of the definitions of the US regulatory system (mentioned under "Biosafety" in this chapter). This set of articles is critical as they clearly define that GMO means any organism including a "new combination of genetic material." To be a GMO, the mere application of "modern biotechnology" techniques is not enough; instead, the donor DNA must come from nonrelated species or species that are not sexually compatible with the receiving species. The insertion must be stable and simultaneous of one or more genes or DNA sequences forming part of a defined genetic construction, so all organisms that do not have these by the "null segregants" of the selection procedure are excluded.

organism did not pass on to the progeny of the organism, so that the nucleic acid of the donor organism did not alter the progeny's sequence.[17]

12. "Modern biotechnology": This means the application of:

   (a) Nucleic-acid in vitro techniques, including recombinant deoxyribonucleic acid (DNA) and the direct injection of nucleic acid in cells or organelles, or
   (b) Cell fusions beyond the taxonomic family, which exceed the natural physiological barriers of reproduction or recombination and which are not techniques used in traditional reproduction and selection.[18]

13. "Biological material": Any plant matter containing self-reproducible or reproducible information in a biological system.[19]

14. "Biotechnological invention": Any biological material obtained via a procedure that is essentially biotechnological.[20]

15. "Essentially biotechnological procedure": Any procedure in which at least one of its steps is a biotechnological technique and not a natural phenomenon such as crossing and selection.[21]

16. "Biodiversity": This includes the variety of organisms in the world and the sets they form, assuming that sets mean ecosystems.[22]

17. "Genetic diversity": This includes all "biodiversity" that may be inherited, including interspecific and intraspecific variation and a large part of diversity in ecosystems.[23]

18. "Plant genetic resources": Any part of "genetic diversity" with current or potential value.[24]

4.1.3 Article 3. Conservation, Sustainable Use, and Fair and Equitable Distribution of Benefits Resulting from the Use of Germplasm

1. To attain the purpose and objectives under Article1.1(a)(b) and in particular the conservation, sustainable use of germplasm, and the fair and equitable distribution of any benefits resulting from their use, as established under Article 1.2(a), every member state, in accordance with domestic laws and in cooperation with other member states when applicable, shall promote an

---

[17] See footnote No. 31.

[18] The definition of "modern biotechnology" has been taken from the Cartagena Protocol.

[19] Usual definition of "biological material."

[20] Usual definition of "biotechnological invention."

[21] Usual definition of "essentially biotechnological invention."

[22] Usual definition of "biodiversity."

[23] Usual definition of "genetic diversity."

[24] Usual definition of "plant genetic resources."

integrated approach regarding the prospection, conservation, and sustainable use of germplasm as defined under Article 2.1.[25]

2. In their relationships with other states, member states recognize the sovereign rights of states over their own germplasm, including that the power to determine the access to these resources corresponds to domestic governments and is subject to domestic legislation.[26]

3. In exercising such sovereign rights, member states agree to establish an open procedure (OP) that is fast, agile, effective, and transparent to facilitate access to germplasm by any individual or entity so requesting and to share, in a fair and equitable manner, the benefits resulting from the use of such resources, based on basic agreement, complementarity, and mutual strengthening.[27]

4. The OP shall include all the germplasm defined under Article 2.1 and which is under the administration and control of member states.[28]

5. Member states agree to take any appropriate measures to promote and facilitate things for individuals and entities within their jurisdiction and which have germplasm to include such resources in the OP.[29]

6. Any request for germplasm access and use cannot be rejected without well-founded grounds. Any conflicts resulting from the access to or use of germplasm shall be decided by the Governing Body.[30]

7. Access to the germplasm by individuals or legal entities of nonmember states shall be determined by the Governing Body.[31]

8. Access shall be exclusively granted with the purposes of use and conservation for research, plant breeding of any type, and technological evolution, and modern

---

[25] The scope of this article must be analyzed together with the scope of the definition of germplasm in Article 2.1. Germplasm includes both tangible structure and intangible information of the genetic resource, so the scope exceeds what has been established in the Nagoya Protocol and the International Treaty on Plant Genetic Resources for Food and Agriculture. This is a *governing principle*.

[26] This article exclusively refers to relationships between states, and the sovereignty each of them has over those resources. It does not mention the obligations of each state stemming from this Treaty, which are explained in the following articles.

[27] The OP (open procedure) is a *governing principle*. While the sovereignty of PGRs remains under the authority of each country, adhering to a treaty entails the immediate facilitation of resources for anyone so requesting them and to participate in the benefits of their use on a complementarity and mutual strengthening basis between the provider of the resource, the one who requests it, and the financial support needed.

[28] There shall be no germplasm that is not covered by the System.

[29] An integral part of the OP is to attain the widest availability of germplasm possible. Given this, it must be promoted and facilitated that all germplasm is accessible for access and use.

[30] Access to and use of germplasm within a member state or among member states is recordable but automatic at the same time. Access to and use of germplasm except for major and fully justified reasons shall not be denied, and the Governing Body will have the final word in these conflicts. The OP and automatic access and use are essential parts of the System.

[31] The cases of access and use requests in connection with germplasm by individuals or entities from nonmember states of the System shall be decided via a procedure to be established by the Governing Body.

biotechnology, provided that such purpose does not entail any chemical, pharmaceutical, or any other industrial applications unrelated with food or animal feed.[32]

9. The OP under Article 3.3 entails that access to germplasm will be granted fast and expeditiously, stating the source of every sample and the receiver of them.[33]

10. With the germplasm provided via the OP available passport data will be provided and, under current regulations, any other nonconfidential associated descriptive information available.[34]

11. Receivers of germplasm via the OP are expressly restrained from claiming an intellectual property or other right limiting the facilitated access to plant germplasm in the received form of the OP.[35]

12. "As received" shall mean the form in which the germplasm was when the access was facilitated via the OP.[36]

13. Access to germplasm protected by intellectual property or other rights, as defined under Article 2.2, shall be consistent with the provisions of this System.[37]

14. Receivers of germplasm via OP and which have been preserved shall remain at the disposal of the OP.[38]

15. In urgent situations due to disasters, member states agree to facilitate the access to germplasm via OP to contribute to reestablishing agricultural systems.[39]

16. Member states recognize: (a) that all the extent of current germplasm comes from a previous germplasm, and that it comes from a prior one, and so on and so forth, and that the access and use facilitated by the germplasm is by itself the main asset of the OP; and (b) that benefits resulting from the access to and use of germplasm must be treated fairly and equitably via the distribution of benefits

[32] Any access and use that is not exclusively for the purposes of plant breeding of any kind and technological evolution, and modern biotechnology are excluded from the scope of the System. This explains the reference to pharmaceutical, chemical, and any other industries.

[33] Access to the resource must be automatic, fast, and expeditious. This is a *governing principle*. Benefits are a substantial part of the System and are governed by Article 3.17.

[34] Administrative procedures of access entail the required documentation so that samples may go around in the country that requested them and among countries.

[35] The application or granting of a title of physical ownership, intellectual property or any other reason on the germplasm as received is fully and entirely affected by nullity. This is a *governing principle*.

[36] "As received" shall be interpreted as the genetic development status of germplasm at the time of facilitated access. Any kind of genetic improvement of the subsequent germplasm at the moment of facilitated access shall not be interpreted to be "as received," unless there is another biological material in the same germplasm and which refers to that status of genetic development progress.

[37] If the germplasm accessed and used pertains to the "protected germplasm" category, it is subject to all considerations incorporated into the different sections of the System.

[38] Germplasm "as received" must be preserved by the one receiving it and must always be available as required by the Governing Body.

[39] In very urgent situations or disasters, germplasm may become an essential survival tool. Under these conditions, germplasm must become available for humanitarian purposes that exceed the purposes and objectives of the System. This is a *governing principle*.

obtained for the previous commercialization of plant varieties, this being the compensation and the second main asset of the OP.[40]

17. To comply with the provisions of section (b) above, member states agree that any seed unit of plant varieties marketed by any individual or legal entity, whether private or government-owned of any member state, shall contribute to the OP 0.5% of the price of that unit before tax.[41]

18. The administrative procedures so that financial resources contributed by the member states be distributed equitably shall be determined by the Governing Body.[42]

19. Member states agree to take steps with the purpose of distributing commercial benefits, via the participation of public and private sectors in certain activities, via partnerships and collaborations between the parties, and with the main purpose of strengthening research promoting and preserving the germplasm's biological diversity, increasing as much as possible intraspecific and interspecific genetic variation to the benefit of farmers and, after complying with these purposes, guaranteeing that any benefits obtained from the facilitated access to germplasm via the OP directly or indirectly go to low-resource farmers of member states, and particularly to native peoples, ethnical and local communities in connection with the conservation and sustainable use of germplasm and any group or community that has particularly made or is making germplasm in the "traditional development" category, as stated under Article 5.[43]

---

[40] With the purpose of reaching the goals of preserving current germplasm and increasing its genetic variability, promoting the conservation and sustainable use of germplasm and stimulating the conservation and sustainable use of PGRs, it is necessary to significantly fund the System globally. For those purposes, the premise was that all protected germplasm must have used a free-use genetic resource at a certain near or far point in time of its development. Therefore, there cannot be any protected germplasm that should not recognize this right that is preexisting to its development and that belonged to mankind as a whole. Fair and equitable distribution of benefits derived from the use of germplasm begins with the funding of the system *ex ante*, that is, when marketing the improved germplasm at any time. This is a *governing principle*.

[41] Considering that estimations state that the global seed market will be approximately 120 billion dollars by 2022 (http://www.marketsandmarkets.com/Market-Reports/seed-market-126130457.html?gclid=CJqvrYC7htQCFckIkQod4T8EqQ), this would mean that 500 million dollars are available a year, applying the contribution rate proposed by this article. This value is more than 20 times higher than the financing goal established by the Governing Body of the Treaty on Plant Genetic Resources for Food and Agriculture for the 5-year period from 2009 to 2014 (ftp://ftp.fao.org/ag/agp/planttreaty/publi/funding_strategy_compilation_en.pdf).

[42] The Governing Body is the unit administrating the funds.

[43] After receiving the funds from the Governing Body, the member states shall apply the resources to the goals of the System.

4.1.4 Article 4. Property Rights

1. To achieve the goal and the purposes of Article 1.1(c)(d) and in particular the property right, as established under Article 1.2(b), member states shall establish an integrated intellectual property protection approach over germplasm as the only and exclusive form of protection, based on a common right and with uniform effects, in which the right which may be granted to a member state shall be valid with the same scope, transmission, termination, and duration as in other member states.[44]
2. Member states shall refrain from granting any other kind of national right over germplasm.[45]
3. The territory of the property right is made up by the territories of multiple member states.[46]
4. Applicants for ownership title may choose any member state to file their first application process.[47]
5. Any new kind of germplasm qualifies to be protected under a property right.[48]
6. Any ideas, principles, processes, and procedures that underlie any element of germplasm cannot be protected via intellectual protection for this Treaty.[49]
7. Under the provisions of this Treaty, five types of intellectual protection are recognized over germplasm under the Treaty:

   (a) Protection of traditional developments
   (b) Protection of plant varieties

---

[44] Territoriality as defined under Article 4 has followed the model established under Articles 1 and 2 of European Council Regulation (EC) No. 2100/94 of 27 July 1994 relative to the community protection of heterogeneous plant varieties. Article 1 provides that "A system of Community plant variety rights is hereby established as the sole and exclusive form of Community industrial property rights for plant varieties," and Article 2: "Community plant variety rights shall have uniform effect within the territory of the Community and may not be granted, transferred or terminated in respect of the abovementioned territory otherwise than on a uniform basis." This is a *governing principle*, which in this case deviates from the TRIPS and UPOV. In these agreements and treaties the following were applied: (a) the "most favored nation" (MFN) principle whereby member states cannot establish commercial discrimination against other member states; if a special commercial condition is given, the same has to be done with all other member states (TRIPS, Article 4), and (b) the "national treatment principle," whereby imported merchandise and goods produced in the country, domestic and foreign services, trademarks or manufacturing marks, copyrights, and domestic and foreign patents must receive the same treatment (TRIPS, Article 3; UPOV 1978, Article 3; UPOV 1991, Article 4).

[45] Along the lines of Article 4.1, but against the European Regulation mentioned, Member States are not allowed to grant any kind of ownership titles over germplasm as defined in the System.

[46] Under Article 4.1, the territory is defined by the addition of member state territories.

[47] Under Articles 4.1 and 4.3, the process to grant an ownership title in the System may be started in any member state.

[48] To qualify as protected germplasm under Article 2.2, any germplasm must be new as per the definition of novelty of every type of intellectual protection. It is implicitly repeated that any natural germplasm may have been "protected germplasm," but not the other way around.

[49] The System exclusively protects products and not the techniques to develop them or any underlying ideas.

(c) Protection of heterogeneous plant varieties
(d) Protection of microorganisms
(e) Protection of biotechnological inventions.[50]

8. Under the definition of Article 2.2, any natural germplasm cannot be subject to the protection types mentioned under Article 4.7. An ownership title over natural germplasm will be incurably null and void.[51]

9. Protections mentioned under Article 4.7 may coexist in the same individual. If such is the case, the coexistence of current protections must be explicit both at the level of official registration and at the marketing level, but any commercial act must be understood as applied to, covering, and transcending a single inseparable entity or individual.[52]

10. Except for Article 5.2, the property right over all other types of protection mentioned under Article 4.7 shall be granted for 12 years, since date of granting, save for any germplasm belonging to fruit or forest species, which shall be 20 years. If a protected germplasm cannot be marketed due to regulatory requirements, the term shall begin since that regulatory authorization has been granted.[53]

---

[50] The purpose is that no plant creation resulting from the application of cultural knowledge, such as science and technology, is left outside protection. There is no prevalence among rights. This is a *governing principle*.

[51] Given the importance of this governing body, we once again remark that no natural germplasm may be protected by any kind of IPRs.

[52] This article is of utmost importance as it expressly establishes the handling and interpretation of rights in the event they coexist in the same organism. This article establishes that: (a) there is the possibility of coexistence of some rights established under Article 4.7 in the same organism; (b) if this is applicable, the coexistence of multiple rights must be explicit, not only at the level official registration, but also at the level of labeling or identifying the commercial product; (c) that any commercial act may be applied to a single indivisible organism; and (d) that coexistence and, therefore, the potential collection on the commercial act is over property rights in force. In this article, the following is implicit: (a) any potential commercial relationships that may exist if it happened that several developers are the holders of rights are decided privately and are not part of the System, and (b) that no property rights that are not in force may be collected on. These are four new *governing principles* for the System.

[53] Except for traditional developments that are not subject to any expiration, the duration of property rights has been set on 12 years since the date when the right was granted, except for fruit and forestry species (given their life and reproduction spans). This introduces a series of concepts departing from current laws. First, for plant varieties the protection time has been significantly reduced, which responds to the observation that life cycles of most commercial varieties that have extensive and intensive crops rarely exceed 10 years of commercial life. The duration of the protection is taken in this case since granting, which is aligned with different UPOV Conventions. The duration of protection on microorganisms and biotechnological inventions (usually protected via patents), has also been significantly reduced, but in this case the lapse is counted from the granting date and not the filling date, so it is possible to think that there will be a substantial difference with the current situation. This way of measuring the lapse of protection departs from what is established under Article 31, TRIPS, setting forth that the lapse is calculated since the request. This seems to be a mistaken solution, as it does not consider the multiple lapses established in domestic regulations with the possibility in many cases to highly affect the interests of legitimate developers, but it harmonizes the entire set of regulations. When regulatory lapses prevent marketing, the date from which protection is calculated is that of commercial approval. This is a *governing principle*.

11. The holder of a property right over protected germplasm will be the individual or group of individuals who have created or, when the regulations of the member state so permits, the entity considered to be the holder of the right by the same regulations.[54]

12. When a protected germplasm has been created by a group of individuals jointly, exclusive rights shall be common property.[55]

13. When a protected germplasm has been created by an employee exercising their tasks or following the instructions of the employer, the employer shall be the only one entitled to exercise any financial rights on the plant germplasm, except for any contractual provision to the contrary.[56]

14. Protection titles established under this Treaty may be requested after another or similar protection title has been requested in a nonmember state. When this is so, the date when the first protection request has been filed shall be considered the priority date, regardless of the time lapsed from the original presentation.[57]

15. When several holders claim title of protection over the same germplasm separately, the right shall be granted to the one who first filed the application for such germplasm.[58]

16. The applicant may withdraw the application at any time during the process.[59]

17. Protection titles defined under Article 4.7 establish the exclusive right to do or authorize the following for the holder:

    (a) Production, reproduction, or temporary or permanent multiplication of the protected germplasm by any means, in any manner, as a whole or its component parts

    (b) Preparation or conditioning with purposes of production, reproduction, or multiplication

    (c) Sale offer

    (d) Sale or any other form of marketing

    (e) Export

    (f) Import

    (g) Possession for any of the purposes under (a)-(f) above

    (h) Exclusively for the category of traditional development germplasm barter and exchange with germplasm of the same category has been added.[60]

---

[54] Usual definition of holding the protection right.

[55] Usual definition of co-ownership.

[56] Usual definition in the event the developer is an employee.

[57] If a nonmember state has granted an intellectual property right that can be assimilated to the ones considered by the System, the priority date is that of the original filing.

[58] Introduction of the "first to file" principle, as opposed to the "first to invent" principle. This is a *governing principle*.

[59] Usual definition in the event that the holder withdraws the application.

[60] Exhaustive definition of multiple commercial acts that are exclusively under the authorization of the holder and which may be done with germplasm. The special situation of traditional developments is considered, as this incorporates two specific marketing acts with the purpose of responding to the special characteristics of this material. This is a *governing principle*.

18. The holder of the right may establish the conditions and limitations of the exclusive right granted under Article 4.17.[61]

19. The holder of the right may assign or transfer by succession the holder's right to protected germplasm and may enter into license contracts, but the holder may never reject a license, applying for this case the principles specified under Article 31 of the TRIPS.[62]

20. It is permitted without prior authorization from the holder to use the crop product that has been obtained after having planted protected germplasm for a new production or reproduction of such germplasm provided that it has been accessed legally and that this new production or reproduction is made in the same initial production unit. If this happens, member states must implement or authorize means so that the holder of the protected germplasm receives reasonable remuneration during its protection period.[63]

21. Any kind of experimentation, development, and/or improvement with any natural or protected germplasm is allowed provided that it has been legally accessed. If the new germplasm obtained after this experimentation is outside the scope of any of the types of protection specified under Article 4.16, that new germplasm may be registered and freely used by the new holder. If the new germplasm is within the scope of protection claims of biological invention or is an essentially derived plant variety, the holder of the right may establish the conditions and limitations of the exclusive right granted under Article 4.17, but an authorization may never be withheld.[64]

---

[61] The scope of the right for the holder includes establishing and limiting commercial acts.

[62] An ownership title over a germplasm may be assigned or transferred, and also licensed, to a third party. In this case, the license is compulsory. This is a *governing principle*. This article refers to Article 31 of the TRIPS, which establishes the conditions for compulsory licenses, to wit: (a) the license applicant must have tried to negotiate, under reasonable terms, a voluntary license with the holder of the property right; (b) the holder of the right has to receive adequate remuneration; (c) the compulsory license shall be nonexclusive, so the holder of the right may make commercial use of the product (this means that the patent holder can continue with production).

[63] For all cases of self-reproducible germplasms, its production or reproduction via natural processes does not require authorization by the holder of the right, but to that end two conditions are compulsory: (a) Legal access to the initial matter, and (b) that the amount produced or reproduced by adequate and sufficient for the same initial production unit (in other words, if self-reproducible germplasm is acquired for initial use in a 100-hectare area, the amount of crop product that can be reserved and used must be equal to or lower than the amount needed to plant a 100-hectare area). As the production or reproduction in these terms and with a natural cycle is an identical copy of the protected material, it is compulsory for member states to implement or authorize a system for holders to receive remuneration, which must meet two requirements: (a) it must be reasonable, and (b) it must last until the right granted by the System is terminated. The definition of "reasonable" is: (a) "appropriate" to the situation of each Member State; (b) "proportional" to the value of the initial germplasm, and (c) "not exaggerated" to the extent that it cannot exceed the initial value. At the same time, this definition entails that it is necessary to be aware of that value in the first commercialization of germplasm. This is a *governing principle*.

[64] A universal exception is established for research and development on all germplasm, which may be done without the holder's authorization. The original idea is part of the conception of the *open source* protection system of UPOV Conventions. But if the product resulting from such research

22. Except for the provisions of Articles 4.20 and 6.8 for the protection of plant varieties, the right of the holder shall not extend to any acts relative to the protective germplasm that has been otherwise sold or marketed by the holder or with the holder's consent, or any material resulting from such material, unless those acts

   (a) entail a new reproduction or multiplication in any way of the protected germplasm in question;
   (b) entail exporting material of the protected germplasm, which allows for its reproduction to a country that does not protect the components of the germplasm, except if the material exported is for consumption.[65]

23. No member state may limit the free exercise of a right of the holder of protected germplasm except for public interest reasons. When such limitation has the effect of allowing a third party to do any of the acts mentioned under Article 4.17, the member state shall adopt all measures required so that the holder of the protected germplasm receives equitable remuneration.[66]

24. The right of the holder may be declared null if it were established that:

   (a) the compulsory conditions established under this Treaty were not effectively met by the time of granting, and
   (b) it was granted to a person who did not have access to it, unless transferred to the person entitled to that right.

   No right of the holder may be annulled by any grounds other than the ones mentioned under this Article.[67]

25. The right of the holder may be declared terminated if it were established that:

   (a) The compulsory conditions established under this Treaty to grant a title are not actually met;
   (b) The holder does not submit before the authorities any information, documents, or materials considered to be necessary to control the maintenance of the protected germplasm; and
   (c) The holder has not paid the due fees to maintain his or her right's validity.

---

and development is within the scope of all or parts of the protected germplasm, the scope of the right for the holder includes establishing and limiting commercial conditions, but licenses are compulsory. This is a *governing principle* and it has been taken from the examples of existing patent laws in Germany, France, the Netherlands, and Switzerland, and the "ISF View on IP" document (ISF 2012).

[65] The right is exhausted after the first marketing instance, except it is a new production or reproduction or is exported to a country that is not a member state and does not protect any germplasms. This article does not cover any such production or reproduction made by the user (Article 4.20) or in the case of plant varieties, if the holder of the right had no reasonable opportunity to exercise his or her rights. This is a *governing principle*.

[66] The principle of "public interest" is introduced as the only grounds to limit the exercise the right of a holder. This is a *governing principle*.

[67] Usual definition of nullity of the right.

Also, the holder's right will terminate in the following cases:

(d) Termination of its validity.
(e) Waiver by the holder. If the right is held by more than one person, the waiver must be jointly made.
(f) Due to not covering the fee mentioned under Article 4.27.

No right of the holder may be terminated by any grounds other than the ones mentioned under this Article.[68]

26. Actions for nullity and termination may be filed as defenses.[69]
27. The applicant for a property right on germplasm shall file:

(a) As far as possible, the germplasm sources used to create the new germplasm
(b) If any of these sources comes from a nonmember state of this System, any evidence showing prior informed consent to the provisions of the Convention on Biological Diversity or via the multilateral system of the International Treaty on Plant Genetic Resources for Food and Agriculture.[70]

28. Except for any products protected under Article 5, the protection titles granted via this System shall pay an annual fee to be established in the pertaining regulation.[71]
29. Any protection titles granted via this System shall be accessible via freely accessible electronic means. The process request shall also be accessible in a summary and this shall not affect the invention's novelty.[72]
30. Any person may challenge with well-founded reasons any protection title granted or to be granted under this System. The challenge process shall be regulated via rules and the purpose of that will be revocation or maintenance of the title.[73]

---

[68] Usual definition of termination of right.

[69] Usual definition of challenge to actions for nullity and termination.

[70] It shall be a requirement to disclose the sources of germplasm, whether natural or protected, which have been used for the purposes of creating new germplasm, avoiding the requirement to disclose the "country of origin" of the resource, as established under the CBD, which is impractical and often impossible to comply with. If the germplasm comes from a nonmember state of the System, all requirements of such international treaties shall be observed. This is a *governing principle*.

[71] An annual fee shall be paid for the maintenance of the registry of rights on germplasm, except for any rights on traditional developments that do not pay any fee.

[72] Transparence is another *governing principle* in the System. For these purposes, a request for protection via IPRs of germplasm shall be accessible in a summarized manner via electronic means, and so will the granting of the right in full. This access must be free for any interested party. Filing the application does not affect the novelty criterion where applicable.

[73] Ways to challenge in many intellectual property regimes provide the opportunity to third parties to challenge the granting during a period, which is not specified by the System. Challenges can also be filed before or after the right was granted. India is an example where the challenge may be filed before or after, and this possibility has been chosen.

31. In consistency with Articles 4.7 and 4.9, all products protected under this System shall have the indication of the pertaining registration number and the type of protection for the germplasm, whether on the product itself, on the label, or on the package. The only exceptions to this requirement shall be any products to which, due to their nature, this requirement cannot be applied. Omitting this requirement does not affect the validity of the protection, but individuals or entities not observing this requirement shall not be entitled to file any criminal actions referred to here. When there is a pending application, that situation must be informed, in the event the products covered by such requests are manufactured, marketed, or imported with commercial purposes.[74]

32. Right holders may resort to civil actions to enforce all intellectual property rights under this System and also criminal procedures and sanctions in other cases of infringement of intellectual property rights, especially in cases where there is a willful purpose to defraud at commercial level.

    Any persons found liable based on infringement by the court will be subject to pay legal costs and damages to the right holder. Any elements or tools used in the commission of any of the crimes mentioned under this article and any illegally produced products will be seized to the benefit of the right holder. Also, the authority hearing the case may immediately order that the products be confiscated, without prejudice to adopting any precautionary relief that was appropriate.

    Recidivism shall be punished with twice the sentence stated in the first paragraph.[75]

33. The property right shall be separate from any measure adopted by member states to regulate, within their territory, the production, certification, and marketing of germplasm materials or the import and export of that material. In any event, these measures shall not affect the application of this System's provisions.[76]

---

[74] This article specifies the contents of the information of labels, bags, and packages under Article 4.9 on the rights of Article 4.7. It is hereby established that the lack thereof does not affect the scope of protection, but it does affect the holder's defense rights. This is a *governing principle* of the System that is based on Swiss patent law.

[75] A civil and criminal sanctioning regime is established in the event the violation is performed with intent to defraud or when it is for commercial purposes. The text of TRIPS Annex 1C, as amended on January 23, 2017, has been considered. As traditional developments are also part of germplasm, for the first time a strong defense of traditional agriculture is made, highly increasing its protection and putting all property rights on an equal footing. This is a *governing principle*.

[76] Usual definition of separation between the System and trade control.

4.1.5 Article 5. Intellectual Protection of Traditional Developments

1. Any of the following may be subject to intellectual protection as traditional developments: groups of plants derived from the application of cultural knowledge, traditional practices and knowledge and innovative concepts of native peoples, ethnic and local communities in connection with the conservation and sustainable use of germplasm, or which has been selected or altered by human action through generations.[77]

2. Traditional developments derived from the application of cultural knowledge, traditional and innovative practices, and knowledge of native peoples, ethnical and local communities in connection with the preservation and sustainable use of germplasm or which has been selected or altered by the action of humans through generations, are the intellectual property of them. These may only be used within and in consistency with the provisions of this System.[78]

---

[77] Article 5.1 defines the scope of what is known here as "traditional developments." Recognizing that a single definition in such a complex issue is not adequate, the WIPO understands "traditional knowledge" as "a living body of knowledge that is developed, sustained and passed on from generation to generation within a community, often forming part of its cultural or spiritual identity," which is made up also by genetic resources and traditional cultural expressions (WIPO 2015). The debate about the protection of what is known here as "traditional developments" is long-standing, and the core disagreement has to do with the attempt to apply individualist regimes as opposed to collective knowledge. The WIPO Intergovernmental Committee on Intellectual Property and Genetic Resources, Traditional Knowledge and Folklore is leading negotiations with the purpose of reaching an agreement on an international legal instrument ensuring the effective protection of traditional knowledge, traditional cultural expressions, and genetic resources. But there seems to be no interaction between this potential instrument with the number of treaties and conventions on IPRs under Chap. 2 of this work, which could lead to increased complex interactions as explained in Chap. 1. That being said, I agree with expert Natalia Franco's view: "The use of intellectual property devices (patents, trademarks, copyrights, appellations of origin, among others) to protect traditional knowledge of native peoples has failed because it has not been considered that these peoples do not share the individualist conceptions of private property held by most Western countries, they do not agree with the existence of temporal protection limits, and they do not like to describe knowledge, as they believe describing is equal to limiting." This author alleges that a protection proposal will only be effective "after collecting and analyzing the information on laws and the traditional practices of these peoples, accepting the sacred and non-rational nature of this knowledge, recognizing the collective property of discoveries, understanding that inventions may be intergenerational in nature (without losing novelty for protection), does not require any registration and is not limited in time." I agree with this author that, within the scope of this article, she attempts to protect "traditional developments" because: "(a) This is an issue of 'justice' or 'equity'; (b) The need to preserve the traditional practices and cultures of native peoples to maintain the natural and biological diversity on the planet; (c) The need to prevent any unauthorized parties to appropriate traditional knowledge; and (d) The importance of traditional knowledge for the social, economic, and cultural development of society at large." (Franco 2007). In connection with "justice," which can be assimilated to "equity," it is necessary to emphasize that this protection must protect "traditional developments" from improper use (Correa 2001).

[78] Article 5.2 recognizes that traditional developments are an intellectual product of their creators and not of the member state, and that they can be protected via an intellectual right. This is a

3. The intellectual protection of traditional developments does not expire.[79]

4. Intellectual rights over traditional developments exist and are recognized with the mere existence of a cultural practice or the cultural knowledge regarding germplasm and does not require prior declaration, express recognition, or official registration, so it may cover any practices that may qualify under such categories in the future.[80]

5. The Governing Body, in coordination with authorities with powers regarding intellectual property in every member state, shall consider all matters relative to intellectual property rights of traditional developments in connection with the preservation or sustainable use of biological diversity, with the purpose of protecting these communities' rights over their knowledge on this matter.[81]

6. To register an intellectual property right over traditional developments granted outside the territorial scope of member states, the Governing Body shall require immediate verification via the pertaining documentation issued by the competent domestic authority in the country where the resource originated.[82]

7. Failure to comply with the above or any falsity or inaccuracy in the information provided by the applicant will result in the denial, suspension, or nullification of the application.[83]

4.1.6 Article 6. Intellectual Protection of Plant Varieties

1. It will be possible to protect via intellectual rights any plant varieties of every botanical gender and species, including any hybrids of genders or species that are:

---

governing principle. However, and for this specific case of right, the exhaustive description of its procedural and administrative phase, together with granting requirements—especially the "novelty" requirement—is reduced to its minimum expression or does not exist altogether, considering the unique nature of the matter to be protected, including the accurate identification of the group or community that produced the traditional development, the time when this took place, and the source of origin. In connection with this, the special feature that the System defines the "territory" as the addition of the territories of member states (Article 4.3) helps to overcome these challenges. As regards the provision that "these may only be used within and in consistency with the provisions of this System," specific reference is made to the fact that they have to be an essential part of what is defined under Article 3.3, and that their contribution shall be expressly recognized under the scope of Article 3.19.

[79] We concur that there are no temporal limits for the protection of traditional developments. This is a governing principle.

[80] We agree that describing a certain knowledge is a kind of limitation to it and that the mere existence of the relationship between cultural practice and traditional development is sufficient proof of that. This is a governing principle.

[81] Another task of the System's Governing Body will be to manage these rights.

[82] A procedure to claim ownership over a traditional development from a state that is not a member of the System is established.

[83] The grounds for denial, suspension, or nullification of an application or grant of a right over traditional developments from a state that is not a member of the System are established.

   (a) Novel
   (b) Distinct
   (c) Uniform
   (d) Stable
   (e) Univocally denominated.[84]

2. A plant variety is novel when, at the date when the application was filed, the components of the variety or the material harvested of that variety have not been sold or assigned to others by the holder of the right, with his or her consent, or to commercially profit from the variety, before these periods:

   (a) One year before the date mentioned above, in the territory of the member states
   (b) Four years or, in the case of trees or vines, six years before the date mentioned above, outside the territory of the member state.[85]

3. A plant variety is distinct if it is possible to differentiate it because of the genotype or genotype combination and because of the expression of at least one qualitative characteristic resulting from a particular genotype or genotype combination, of any other plant variety whose existence is evident by the date the application is filed. The existence of another variety will be considered evident especially if at the date when the application was filed:

   (a) Such variety has been registered with an official registry of plant varieties, in any state, or in any intergovernmental organization with authority for that purpose
   (b) An application has already been filed for protection or for registration in an official registry of varieties, provided that such application has resulted in granting or registration.[86]

4. A plant variety is uniform when, setting apart the variations that can be expected from the specific characteristics of its propagation, it is uniform enough in the expression of its characteristics included in the analysis of its distinctive nature as well as in any others used in the variety description.[87]

5. A plant variety is stable when the expression of its characteristics included in the analysis of its distinctive nature, as well as any others used in the description

---

[84] This entire article builds upon the standard provisions and general requirements for protection stemming from UPOV Conventions. The changes needed to accommodate these principles of the System are included in footnotes. This is a *governing principle*.

[85] General requirements of the "novelty" criterion resulting from the UPOV Convention.

[86] General requirements of the "distinctness" criterion. It is necessary to interpret this criterion together with the definition of "Plant Variety" of Article 2, 4. A "plant variety" is defined by the genotype and the phenotypical expression of the resulting characters of such genotype or of a certain combination of genotypes, and distinguished from any other set of plants for having a new genotype or a certain combination of genotypes and for the phenotypical expression of at least a qualitative character of such genotype or combination of genotypes.

[87] General requirements of the "uniformity" criterion resulting from the UPOV Convention.

of the variety, suffer no alternation after repeated propagation or, when a specific propagation cycle takes place, by the end of any such cycles.[88]

6. A plant variety has a univocal denomination when this designation:

   (a) Is a generic designation
   (b) Is different from any denomination designating a plant variety of the same species or of a related species in any country where there is an official record of plant varieties.[89]

7. The mere discovery of a plant variety whose reproduction or multiplication permits to convert it in a variety does not qualify as a new plant variety.[90]

8. The exclusive rights of the holder of protected germplasm under Article 4.17 shall be applied to the product of the crop if it has been obtained via the nonauthorized use of components of the protected plant variety, provided that the holder has not had a reasonable opportunity to exercise his or her rights over such components of the variety.[91]

9. The exclusive rights of the holder of protected germplasm defined in Article 4.17 and, if applicable, the rights on the product of the crop established under Article 6.8 shall also be applied to:

   (a) Plant varieties that are essentially derived from the plant variety subject to protection, provided that it is not in turn an essentially derived plant variety
   (b) Plant varieties whose production requires the repeated use of the plant variety protected.[92]

10. For the purposes of Article 6.9(a), a plant variety shall be considered essentially derived from another plant variety (the "initial variety"), when:

   (a) It is derived from an initial variety or several subsequent varieties that in turn derive mainly from an initial variety

---

[88] General requirements of the "stability" criterion resulting from the UPOV Convention.

[89] General requirements for the denomination of varieties resulting from the UPOV Convention.

[90] The definition of "breeder" under Article 1(iv) of the 1991 UPOV Act reads: "the person who bred, or discovered and developed, a variety." This definition has been assimilated in multiple domestic legislations for the protection of plant varieties and has created significant confusion, so much so that many texts highlight that one of the distinctive features of breeder rights, as opposed to patents, is that breeder rights permit the protection of discoveries. This is not correct, as the UPOV Convention clearly states that the potential "discovery" must always be accompanied by "development." Apart from this, the underlying assumption in the System is that nothing that is preexisting in nature may be protected, which requires more and explicit clarity to that respect. Based on these reasons, the "discovery and development" requirement has been discarded as criterion for protection. This is a *governing principle*.

[91] Exhausting the first commercialization under Article 4.22 has the exception included in this article. This principle is optative in the 1978 UPOV Act and it is a compulsory principle in the 1991 UPOV Act. That principle is important to apply certain procedures to observe the law. This is a *governing principle*.

[92] Extent of the scope of the right under the UPOV Acts of 1978 and 1991.

(b) It is clearly differentiated from the initial variety
(c) It is consistent with the initial variety in the expression of essential characters resulting from the genotype or the genotype combination of the initial variety.[93]

11. Other regional or domestic regulations adopted by virtue of the previous article may specify potential derivation operations.[94]
12. The application to register a plant variety shall include a declaration about whether or not the new variety is an essentially derived plant variety. Any disputes between the parties shall be decided within the Governing Body.[95]

---

[93] This definition of essentially derived variety contains the full "spirit" of the definition included in Article 14.5(b) of the UPOV Convention of 1991, but it has been modified. The experience of a large number of domestic legislations that adopted the concept of essentially derived variety, but which have often modified it in substantial terms, shows that it is not clear to be understood by all audiences. The original concept requires additional clarification, simplification, and elimination of certain ambiguities and inconsistencies among its three components, particularly between (a) and (c). Section (a) was modified changing the original expression "or of a variety which in turn derives from the initial variety," to "or of subsequent varieties which in turn derive from the initial variety," so that it is perfectly established that there exists or may exist a genetic derivation cascade in all cases based on a single initial variation, which is not conclusively clear in the original version. The expression "preserving at the same time the expressions of the essential characteristics which result from the genotype or the combination of genotypes of the initial variety," was eliminated, as section (a) must be exclusively related to genotypic derivation and not phenotypical conformity under section (c). Section (b) was not modified. In section (c), the phrase "except in connection with the differences resulting from derivation" was eliminated, as this section (c) must be exclusively related to phenotypical conformity and not to the genotypic derivation under section (a). The language of the concept of essentially derived variety in the project is now clearer, simpler, and more specific, as every section (a, b, c) responds to each of the three derivation components: (a) genotypic derivation based on the initial variety or the subsequent derived varieties; (b) the clear differentiation in the loose sense of the UPOV Convention, and (c) the phenotypical conformity between the essentially derived variety and the initial variety. Therefore, it is explicit that a derived variety: (1) genotypically derives directly from an initial variation or from subsequent derived varieties (section a); (2) must be distinguished at least based on a qualitative character of it (section b); and (3) its phenotype is consistent with the initial variety (section c).
The six requirements of essential derivation are defined by Articles 9.9 and 6.10. The three technical requirements establishing the derivation are: (1) clear distinction of the initial variety (6.10(b)); (2) phenotypical conformity (6.10(c)); and (3) derivation based on the initial variety or derived consequents (6.10(a)). The three legal requirements establishing dependence are: (4) the initial variety must be protected (6.9(a)); (5) the initial variety cannot be at the same time an essentially derived variety (6.9(a)); and (6) dependence can only exist in favor of a single initial variety (6.10(a)) (Rapela 2006, 2008; ISF 2012). This is a *governing principle*.
[94] Article 14.5(b)(c) of the UPOV Act of 1991 lists a nonexhaustive detail of any acts that may be considered determinant for derivation. The choice was made not to mention them given the relentless progress and news in connection with plant breeding and genetics, and let them be specified in regional or domestic regulations.
[95] This requirement departs from UPOV 1991, where the burden of the proof rests with the side that had allegedly initiated the derivation. Here, conversely, it is provided that the one who derived it must report it, leaving the developer of the alleged initial variety the power to claim for a potential falsity in the application. This is a *governing principle*.

4.1.7   Article 7. Protection of Heterogeneous Plant Varieties (HPVs)

1. It will be possible to protect via intellectual rights any heterogeneous plant varieties of every botanical gender and species that are:

   (a) Novel
   (b) Distinct
   (c) Univocally denominated.[96]

2. An HPV is novel when, at the date when the application was filed, the components of the HPV or the material harvested of that HPV have not been sold or assigned to others by the holder of the right, with his or her consent, or to commercially profit from the HPV, before the following periods:

   (a) One year before the date mentioned above, in the territory of the member states of the Governing Body
   (b) Four years or, in the case of trees or vines, six years before the date mentioned above, outside the territory of the member states of the Governing Body.[97]

3. An HPV is distinct if it is possible to clearly differentiate it by the expression of the characteristics resulting from a genotype from any other HPV whose existence is evident by the date the application was filed. The existence of another HPV will be considered evident especially if at the date when the application was filed:

   (a) such HPV has been registered with an official registry, germplasm bank, or publication in any state, or in any intergovernmental organization with authority for that purpose;

---

[96] Article 7 of the System is intended to cure an inconsistency in the TRIPS. Section 5, Patents, Article 27, Patentable Subject Matter, (1) of the TRIPS reads: "patents shall be available for any inventions, whether products or processes, in all fields of technology . . . patents shall be available and patent rights enjoyable without discrimination as to . . . the field of technology." Section 3(b) establishes that "plants" may be excluded from patentability, but plant varieties must be protected "either by patents or by an effective *sui generis* system or by any combination thereof." It is absolutely clear in the TRIPS that member states have flexibility to define the *type* of protection over plant varieties (breeder rights, patents, or a combination of them), but not to define their lack of protection. Instead, all plant varieties must be protected. But it is difficult to imagine that the TRIPS intended to leave some kind of heterogeneous plant variety without protection. As in many domestic systems heterogeneous varieties cannot be protected by patents, a protection gap has been generated on heterogeneous plant varieties that are not plant varieties, that is, sets of plants that are not uniform and stable. As there are very good contributions to worldwide plant breeding of populations, compounds, landraces, and other developments that are not plant varieties, this article pretends to close that gap. This is a *governing principle* of the System, aligned with the provisions of the TRIPS. One of the substantial differences in the protection of the "heterogeneous plant variety" as opposed to the protection of the "plant variety" is that HPVs do not qualify for the application of the concept of "essentially derived variety."

[97] In consistency with the requirement of Article 6.2 on Plant Varieties.

(b) an application has already been filed for protection or for registration in an official registry, catalog, germplasm bank, or publication, provided that such application has resulted in granting or registration.[98]

4. An HPV has a univocal denomination when this designation:

(a) is a generic designation;
(b) is different from any denomination designating an HPV of the same species or of a related species in any country where there is an official registry, catalog, germplasm bank, or publication.[99]

5. The mere discovery of a new HPV whose reproduction or multiplication permits to convert it in a variety does not qualify as a new HPV.[100]

6. The exclusive rights of the holder of protected germplasm under Article 4.17 shall be applied to the product of the crop if it has been obtained via the nonauthorized use of components of the protected HPV, provided that the holder has not had a reasonable opportunity to exercise his or her rights over such components of the variety.[101]

4.1.8 Article 8. Intellectual Protection of Biotechnological Inventions

1. A biotechnological invention may be protected by intellectual property rights when it:

(a) is novel;
(b) entails an inventive activity; and
(c) may have an industrial application.[102]

2. A biotechnological invention is novel when it is not included in the state-of-the-art. State-of-the-art means the set of technical knowledge that has been disclosed to the public before the date of the application for protection or, if applicable, the priority recognized, by an oral or written description, by the use or any other means for disclosure or information, in the country or abroad. A biotechnological invention shall be novel even if it is about a product made up by or containing biological matter isolated from its natural environment or when it already exists in natural state.[103]

---

[98] In consistency with the requirement of Article 6.3 on Plant Varieties.

[99] In consistency with the requirement of Article 6.6 on Plant Varieties.

[100] In consistency with the requirement of Article 6.7 on Plant Varieties.

[101] In consistency with the requirement of Article 6.8 on Plant Varieties.

[102] All of this article is based on the general doctrine of patents. The changes to adapt the doctrine to the System are mentioned in the pertaining footnotes, highlighting the scope and description of the living matter that may be protected. This is a *governing principle*.

[103] General requirements of the "novelty" criterion.

3. A biotechnological invention entails an inventive activity when the creative process or its results do not result from the state-of-the-art in an evident manner for a person with standard knowledge about the pertaining technical matter.[104]

4. A biotechnological invention may be applied in the industry when the creation leads to obtaining a result or an industrial product, understanding that "industry" includes agriculture, forestry, and transformative industries proper.[105]

5. The disclosure of a biotechnological invention shall not affect its novelty when, one year before filing the application for protection or, if appropriate, the recognized priority, the holder or the legal entities defined under Article 4.16 have disclosed the invention by any means of communication or have disclosed it at a domestic or international exhibition. When filing the pertaining application, supporting documents shall be included.[106]

6. Without prejudice to any of the provisions of Article 4.8, a biotechnological invention is:

   (a) A partial or full complementary DNA sequence
   (b) A DNA sequence, a gene, or an element isolated from its natural environment through a technical procedure of identification, purification, characterization, and multiplication and which has also been altered by any mutagenic procedure, either physical (radiation), chemical (mutagenic agents), or site-specific (resulting from any gene-editing technique)
   (c) A DNA sequence or a gene obtained by modern described or to-be-described biotechnological techniques and which in all cases are not exact copies of any DNA sequence or gene that is preexisting in nature
   (d) A new combination of genetic material or transformation event.[107]

7. The following does not qualify to be biotechnological invention:

   (a) Nothing that is preexisting in nature
   (b) Any new germplasm that may be protected as traditional development, a plant variety, an heterogeneous plant variety, or a microorganism

---

[104] General requirements of the "inventive activity" criterion.

[105] General requirements of the "industrial application" criterion.

[106] General requirements of the "loss of novelty" criterion.

[107] One of the defects in some patent laws is the nonexistent, scarce, or ambiguous definition of what is a living matter that may be subject to protection. In this regard, the definition in the System is exhaustive and explicit, following the law of nature doctrine, whereby anything that is subject to natural laws cannot be patentable. In the System, the traditional doctrine of patent is replaced and exclusively limited to "biotechnological invention," which includes: (a) A partial or full complementary DNA sequence, which is artificial DNA obtained based on a messenger RNA mold containing only the sequences codifying proteins and not intron sequences (Source: Association of Molecular Pathology v Myriad Genetics; Barraclough 2013; Egelie et al. 2016); (b) A DNA sequence or a gene or an element isolated from its natural environment but which must have been altered by any mutagenic procedure; (c) A DNA sequence or a gene obtained via modern biotechnological techniques, but which cannot be exact copies of any DNA or gene that is preexisting in nature; and (d) A new combination of genetic material or transformation event, as per the definition of Article 2.11. This is a *governing principle*.

(c) The mere discovery of a partial or full sequence of a gene, including its isolation from its natural environment via a technical procedure of identification, purification, characterization, and multiplication, provided that the structure of such element is identical to the structure of an element in natural state

(d) Any mutant genes obtained by spontaneous mutation

(e) Any inventions whose commercial use should be restricted to protect public order or morality, health, or the life of people or animals or to preserve any vegetables or avoid scientifically proven damage to the environment.[108]

8. In addition, to be a biotechnological invention, the industrial application of a partial or full gene sequence must be expressly included in the application for protection. A mere partial or full gene sequence with no indication of biological function is not a biotechnological invention.[109]

9. The protection granted to a biotechnological invention shall not extend on any biological material, plant variety, or heterogeneous plant variety in which it has been inserted by any natural or essentially biological procedure.[110]

10. The protection granted to a biotechnological invention shall not expire if the biological material, plant variety, or heterogeneous plant variety containing it is used afterward for new reproductions or multiplications.[111]

---

[108] Article 4.6 supplements Article 4.5 and specifies what is not a biotechnological invention and, therefore, does not have any protection within the System: (a) Nothing preexisting in nature, ratifying the *governing principle* under Article 2.3 insofar as this is natural germplasm that cannot be protected; (b) Any new germplasm that may be protected otherwise in the System; (c) The mere discovery of a partial or full sequence of a gene, including any such isolated gene, ratifying that these are products of nature; (d) Any spontaneous mutant gene; (e) Any inventions whose commercial use should be restricted to protect other matters (classical principle of the doctrine of patents). This is a *governing principle*.

[109] An additional requirement to be a "biotechnological invention" is the description of the "industrial application" of the DNA sequence.

[110] This article revisits the issue of coexistence of rights over the same organism mentioned under Article 4.9, which must be interpreted in conjunction. Unlike European regulations (Article 5 of Directive 98/44/CE), in the System rights can coexist in the same organism, but property rights over a "biotechnological invention" do not extend to the organism containing it, which is a unit for commercial purposes. Example: (a) In the case of a variety protected as "plant variety" and which at the same time contains a protection as "biotechnological invention," both rights are valid, may be collected on, but the commercial act is only one; (b) In the case of a variety with no protection as "plant variety" (given that its title expired) and which contains a protection as a "biotechnological invention," the only valid right is the latter—it can be collected on, but the commercial act is only for the biotechnological invention, that is, the right does not extend to the variety. In this case, the plant variety is not protected, but the "biotechnological invention" is. This is a *governing principle*.

[111] The right over the "biotechnological invention" cannot be exhausted if the organism containing it can be produced or reproduced. Every act of production or reproduction of an organism containing a protected "biotechnological invention" entails having done an exact copy of such invention, which remains protected for the term of the right.

11. The protection granted to a biotechnological invention shall not include more than one invention or set of inventions related among themselves so that they are part of a single general inventive concept. The biotechnological invention must be clearly described in the application in clear and complete terms so that a subject-matter expert may apply it.[112]

12. The claims define the object for which protection is sought. The claims must be clear and concise and have to be based on the description mentioned under the previous Article. The right granted shall be determined by the first claim approved, which defines the biotechnological invention and limits the scope of the right. If clarity and understanding so require, the first claim can be followed by one or several claims, known as "dependent claims," aimed at clarifying definitions included in it. The claim can never reach a partial or full sequence of a gene that is preexisting in nature, which has been the basis for the new development.[113]

4.1.9 Article 9. Intellectual Protection of Microorganisms

1. Any microorganisms pertaining to the domains of Archaea, Bacteria, and kingdoms Protista or Fungi, may be protected via intellectual property rights provided that they:

   (a) are novel;
   (b) entail an inventive activity; and
   (c) may have an industrial application.[114]

2. A microorganism is novel when its development is not included in the state of the art or when it is not preexisting in nature. State of the art means the set of technical knowledge which has been disclosed to the public before the date of the application for protection or, if applicable, the priority recognized, by an oral or written description, by the use or any other means for disclosure or information, in the country or abroad.[115]

---

[112] The detailed, clear, and accurate description of the "biotechnological invention" allowing an expert on the subject matter to understand the invention claimed is one of the main components that will be considered to assess whether or not the right will be granted. The description must cover a single invention or a set of very related inventions.

[113] The claims define the scope of the invention, and the first claim must identify the "biotechnological invention." The first claim may be followed by dependent claims with the only purpose of clarifying the scope of the protection with the highest accuracy possible. In no event may a dependent claim contain the natural gene that could have been potentially mentioned in the first claim.

[114] In consistency with the considerations for Article 7.1 and the protection provided for by the TRIPS, microorganisms must be protected. However, unlike HPVs under Article 7, microorganisms are generally considered in the protection of almost all patent laws, which has been replicated here. The article is consistent with the previous article. Article 8.1 is consistent with Article 7.1 on "biotechnological inventions." This is a *governing principle*.

[115] In consistency with the requirement of Article 7.2 on biotechnological inventions.

3. A microorganism entails an inventive activity when the creative process that led to its development does not result from the state-of-the-art in an evident manner for a person with standard knowledge about the pertaining technical matter.[116]

4. A microorganism may be applied in the industry when the creation leads to obtaining an industrial product, understanding that "industry" includes agriculture, forestry, and transformative industries proper.[117]

5. The disclosure of a microorganism shall not affect its novelty when, one year before filing the application for protection or, if appropriate, the recognized priority, the holder or the legal entities defined under Article 4.16 have disclosed the invention by any means of communication or have disclosed it at a domestic or international exhibition. When filing the pertaining application, supporting documents shall be included.[118]

6. Any discovery of a microorganism in its natural state or any microorganisms whose commercial use should be restricted to protect public order or morality, health, or the life of people or animals or to preserve any vegetables or avoid scientifically proven damage to the environment do not qualify as new microorganisms.[119]

7. A microorganism obtained via modern biotechnology qualifies as a new microorganism.[120]

8. To be a new microorganism, the industrial application of its function must be expressly included in the application for protection. A microorganism without any indication of biological function is not a new microorganism.[121]

9. The protection granted to a new microorganism shall not include more than one invention or set of inventions related among themselves so that they are part of a single general inventive concept. The new microorganism must be described in the application in clear and complete terms so that a subject-matter expert may apply it.[122]

10. If the new microorganism is not available for the public and cannot be described in the application for protection in adequately clear and complete terms so that an expert may reproduce the invention, or if it entails the use of such matter, the description shall only be considered sufficient if:

    (a) The new microorganism has been deposited no later than the date when the patent application was filed at an institution recognized for the application. The international depositary institutions recognized as such under Article 7 of the Budapest Treaty, dated April 28, 1977 (Budapest Treaty on the

---

[116] In consistency with the requirement of Article 7.3 on biotechnological inventions.

[117] In consistency with the requirement of Article 7.4 on biotechnological inventions.

[118] In consistency with the requirement of Article 7.5 on biotechnological inventions.

[119] In consistency with the requirement of Article 7.7 on biotechnological inventions.

[120] It is repeated that any natural germplasm, as it is preexisting in nature, cannot be protected via IPRs. The "new" microorganism must be a product of modern biotechnology.

[121] In consistency with the requirement of Article 7.8 on biotechnological inventions.

[122] In consistency with the requirement of Article 7.11 on biotechnological inventions.

International Recognition of the Deposit of Microorganisms for the Purposes of Patent Procedure) shall be considered recognized, without prejudice to adding others.

(b) The application filed contains the relevant information that the applicant has in connection with the characteristics of the new microorganism deposited.

(c) The application for protection mentions the depositary institution and its number.[123]

11. The public shall have access to the microorganisms at the depositary institution as from the date of deposit, as per the conditions established in the pertaining regulations.[124]

12. Access to the new microorganism and the delivery of the sample will only be made if the applicant undertakes not to do any of the following while the effects of the protection last:

(a) To provide any sample of the new microorganism deposited or of a matter deriving therefrom to others

(b) To use any sample of the biological matter deposited or a matter derived therefrom, except for experimental purposes, unless the applicant or the holder of the patent waives such commitment.[125]

4.1.10 Article 10. Biosafety

1. To reach the purposes of Article 1.1(e) and particularly the biosafety of the new germplasm, as established under Article 1.2(c), member states shall see to it that the development, handling, transportation, use, transfer, and commercial release of any living organism, whether or not genetically modified, containing new characteristics, be in such a way to avoid or reduce any risks for biological diversity, the environment, and human and animal health.[126]

---

[123] The deposit procedure for microorganisms is standard practice in most current legislations dealing with the protection of microorganisms.

[124] The same as above.

[125] The same as above.

[126] As explained in Chap. 2, in adopting biosafety regimes to be applied on new biotechnological products, countries have adopted multiple criteria. Beyond the scope they have had, these criteria have resulted in asymmetry and asynchrony in terms of commercial approvals, with deep business impact. Countries that have adopted biosafety regimes based on the process (most of them) apply in practice a binary regulation regime differentiating genetically modified products from products derived from conventional technologies. In light of more than two decades of experience, this regulatory decision seems to be lacking in support, which becomes even more evident with the emergence of NBTs, which had been foreseen a long ago (Rapela 2005). The Canadian biosafety system, which is strictly based on the product, identifies any plants in which one or more characteristics have been intentionally introduced, and the regulation procedure is triggered only when the characteristic introduced is new for the harvested populations of the species and when it has the potential of affecting the specific use and safety of the plant with respect to the environment and

2. For that purpose, member states shall adopt a biosafety assessment criterion over the living organism, whether or not genetically modified, based on the product, regardless of the process whereby it was developed.[127]

3. A "novel characteristic" is any characteristic that is new for stable populations of plant species and which has potential to have an adverse impact on biological diversity, the environment, and human and animal health.[128]

4. Member states shall cooperate among themselves in connection with biosafety and shall promote harmonious frameworks exclusively based on scientific knowledge published in specialized reviews.[129]

5. Risk perception shall not be a judgment element in adopting any biosafety measures.[130]

4.1.11 Article 11. Governing Body[131]

1. Every member state shall consolidate the administration of this treaty at the domestic level in connection with the purpose stated in Article 1.1.[132]

---

the human being. In this system by product, a plant with a novel characteristic is a plant containing a new characteristic for the environment and which has the potential of affecting the specific use and the safety of the plant in connection with the environment and human health. These characteristics may be introduced and be the result of the empirical breeding made by communities, conventional modern breeding, breeding with molecular markers or bioengineering. The System understands that the rationality of the Canadian product-based regulatory approach is higher than the rationale of the process-based approach and is the approach adopted. This is a *governing principle.*

[127] The content of Article 10.1 is ratified—the characteristic of the product, instead of the development or process whereby it was obtained, is what triggers the regulatory process. A new plant selected by a local community based on ancestral or empirical knowledge with a novel characteristic may be regulated, just like if it was obtained via the newest bioengineering technique. Conversely, any kind of product, even a product developed via modern biotechnology, which does not include any novel characteristic, shall not be regulated as it is understood that it already has years, decades, or centuries of safe use. Therefore, the interaction between this article and Article 2, especially with Article 2(10), reveals the standard relationship between regulation and genetically modified living organism. A GMO may or may not be subject to regulation depending on whether or not it has a novel characteristic. This is a *governing principle.*

[128] The definition of "novel characteristic" from the Canadian regulatory system is adopted.

[129] The article introduces two concepts: (a) harmonization of regulation among member states for the purposes of avoiding asymmetry or asynchrony; and (b) the assessment must be strictly based on scientific criteria.

[130] This reaffirms the content of Article 10.4, to the extent that risk perception is not a judgment element in regulatory matters.

[131] This regulates the operation of the System's Governing Body.

[132] This article establishes the consolidation of the domestic administration that separately covered the areas of intellectual property of plant varieties, biotechnological inventions, genetic resources, and biosafety, and which now comes under the scope of the System in a single integrated unit. This is a *governing principle.*

2. For the purposes of Article 1.6, the Governing Body of this System is created, which is made up by the representations of the offices of member states as per the provisions of the above article.[133]
3. All decisions by the Governing Body shall be adopted by consensus, unless another method is agreed upon to reach a decision on certain measures.[134]
4. The tasks of the Governing Body shall be the promotion of the full application of this Treaty, considering its objectives, and in particular:

   (a) to give instructions and guidance on policies for the surveillance and approval of any necessary recommendations to apply this Treaty;
   (b) to approve plans and schedules to apply this Treaty;
   (c) to approve in its first meeting and to periodically assess the financing strategy to apply this Treaty;
   (d) to approve the budget for this Treaty;
   (e) to study the possibility of any auxiliary bodies that may be necessary and their pertaining terms and makeup;
   (f) to establish, if necessary, an appropriate mechanism, such as a fiduciary account, to receive any financial resources to be deposited in it for the amount from the application of Article 3.15;
   (g) to establish any necessary measures with the purpose of distributing commercial benefits as established under Article 3.16;
   (h) to establish and maintain the cooperation with other international organizations and relevant treaty bodies on issues covered by this Treaty;
   (i) to perform any other functions that may be necessary to accomplish the goals of this Treaty.[135]

5. Every member state shall have one vote and may be represented at the meetings of the Governing Body by only one delegate, who may be accompanied by an alternate, and experts and assistants. Alternates, experts, and assistants may participate in the deliberations of the Governing Body, but they will have no vote, except they are duly authorized to substitute the delegate.[136]
6. The United Nations and its specialized agencies, the FAO, the WIPO, and its specialized agencies, as well as any state that is not a member state, may be represented as observers at the meetings of the Governing Body. Any other body or agency, whether governmental or nongovernmental, which is qualified in areas relative to the preservation and sustainable use of PGRs for food and agriculture and which has communicated its intent to the president to be represented as an observer at a meeting of the Governing Body may be admitted. The

---

[133] All the integrated offices of the previous article are the ones with representation in the System's Governing Body.

[134] The purpose of consensus is to be the main method for the System's Governing Body to make decisions.

[135] The tasks of System's Governing Body are established.

[136] Unlike other treaties and conventions in force, the principle is the democratization of decisions, in the event there is no consensus, assigning one vote to each representation. This is a *governing principle*.

admission and participation of observers shall be subject to the internal rules approved by the Governing Body.[137]

7. The Governing Body shall approve and modify, if necessary, its rules and its financial standards, which shall not be incompatible with this Treaty.[138]

8. The presence of delegates representing 50% of member states shall be necessary to reach a quorum at any meeting of the Governing Body.[139]

9. The Governing Body shall hold regular meetings at least once a year.[140]

10. Special meetings of the Governing Body shall be held at any other time considered necessary, whether by an agreed-upon decision or following a request of a member state, provided that such request has the support of at least one-third of member states.[141]

11. The Governing Body shall elect its president, vice president, secretary, and treasurer as per its rules.[142]

4.1.12 Article 12. President

1. The president of the Governing Body shall be elected by member states and shall be assisted by any personnel considered necessary.[143]

2. The president shall have the following functions:

   (a) Organizing meetings of the Governing Body and any of its auxiliary bodies that may be established and give them administrative support
   (b) Assisting the Governing Body in the performance of its functions, especially the performance of specific tasks that the Governing Body may assign to the president
   (c) Reporting on its activities to the Governing Body.[144]

3. The president shall communicate to all member states:

   (a) The decisions of the Governing Body within 60 days since they were approved
   (b) The information received from member states as per the provisions of this Treaty.[145]

---

[137] Conditions are established to be an observer member of the System's Governing Body.

[138] Conditions are established to draft the internal rules of the System's Governing Body.

[139] Quorum conditions are established for the operation of the System's Governing Body.

[140] The regularity of the meetings of the System's Governing Body is established.

[141] Special meetings will be possible and the mechanism to call meetings of the System's Governing Body is established.

[142] The method to elect the authorities of the System's Governing Body is established.

[143] The method to elect the president of the System's Governing Body is established.

[144] The tasks of the System's Governing Body are established.

[145] The communication methods to the president of the System's Governing Body are established.

4. The president shall provide the documents in the official languages of the World Trade Organization.[146]
5. The president shall cooperate with other organizations and treaty bodies connected with this Treaty.[147]

4.1.13 Article 13. Observance[148]

In its first meeting, the Governing Body shall examine and approve effective cooperation procedures and operational mechanisms to promote the observance of this System and to address any violations. These procedures and mechanisms shall include, if necessary, control and offer to assist, including any legal matters.

4.1.14 Article 14. Dispute Resolution[149]

1. If there is a dispute in connection with the interpretation or application of this System, the interested parties will try to resolve it via negotiation.
2. If the interested parties cannot reach an agreement via negotiation, Article 11.3 shall be applied.

4.1.15 Article 15. Amendments to the System[150]

1. Any member state may propose amendments to this System provided that any such amendments do not affect the objectives of the System established under Article 1.
2. Any amendments to the System shall be approved at a meeting of the Governing Body. The president shall communicate the text of the amendment to member states at least six months before the meeting where the approval is proposed.
3. Any amendments to the System shall be exclusively approved by consensus of the member states in attendance at the meeting of the Governing Body, and the dispute resolution method established under Article 14.2 shall not be valid.

---

[146] The official languages of the System's Governing Body are Spanish, French, and English.

[147] The general cooperation guidelines for the president with other organizations are established.

[148] The general procedures to observe the System are established.

[149] The procedures to decide disputes within the System are established.

[150] The procedures to incorporate amendments to the System are established. Any amendment must comply with two requirements: (a) it cannot alter the objectives of the System; and (2) it can only be applied by a unanimous decision of member states.

4. The amendments approved by the Governing Body shall become effective on the 90th day after the date when instruments of ratification, acceptance, or approval by two-thirds of member states have been deposited.

4.1.16 Article 16. Execution[151]

This System will be open for execution from ____ until _____.

4.1.17 Article 17. Adherence

The System shall be open to the adherence by all members of the United Nations.

4.1.18 Article 18. Entry into Force

1. This System shall enter into force on the 90th day after the date when the fourth instrument of ratification, acceptance, approval, or adherence has been received.

4.1.19 Article 19. Reservations

No reservations to this System may be made.

4.1.20 Article 20. Denunciation

2. At any time after two years since the entry into force of this System for a member state, the member state may notify the Governing Body in writing of the member state's denunciation of this System. The Governing Body shall immediately inform all member states.
3. The denunciation shall be effective one year after the date when the notice has been received.

---

[151]Articles 16, 17, 18, 19, 20, and 21 are procedural in nature.

4.1.21 Article 21. Termination

1. This System shall be automatically terminated when, as a result of denunciations, the number of member states is lower than 4, unless the remaining member states unanimously decide otherwise.
2. The Governing Body shall inform all the contracting parties when the number of contracting parties is lower than 4.
3. In the event of termination, assets shall be sold as established in the financial rules approved by the governing body.

References

Adl SM, Simpson AGB, Farmer MA, Andersen RA, Anderson OR, Barta JR, Bowser SS, Brugerolle G, Fensome RA, Frederico S, James TY, Karpov S, Kugrens P, Krug J, Lane CE, Lewis LA, Lodge J, Lymm DH, Mann DG, McCourt RM, Mendoza L, Moestrup O, Mozley-Standridge SE, Nerad TA, Shearer CA, Smirnov AV, Spiegel FW, Taylor MFJR (2005) The new higher-level classification of eukaryotes with emphasis on the taxonomy of protists. J Eukaryot Microbiol 52:399–451. https://doi.org/10.1111/j.1550-7408.2005.00053.x

Adl SM, Simpson AGB, Lane CE, Lukeš J, Bass D, Bowser SS, Brown MW, Burki F, Dunthorn M, Hampl V, Heiss A, Hoppenrath M, Lara E, le Gall L, Lynn DH, McManus H, Mitchell EAD, Mozley-Stanridge SE, Parfrey LW, Pawlowski J, Rueckert S, Shadwick L, Schoch CL, Smirnov A, Spiegel FW (2012) The revised classification of eukaryotes. J Eukaryot Microbiol 59:429–514

Barraclough E (2013) What Myriad means for biotech. WIPO Magazine, 4/2013

CDB (2016) Conference of the Parties to the Convention on Biological Diversity, Thirteenth meeting Cancun, Mexico, 4–17 December 2016. Agenda item 17. Decision adopted by the Conference of the Parties to the Convention on Biological Diversity. Digital sequence information on genetic resources. CBD/COP/DEC/XIII/16

Correa CM (2001) Traditional knowledge and intellectual property: issues and options surrounding the protection of traditional knowledge: a discussion paper. Quaker United Nations Office, Geneva

Egelie KJ, Graff GD, Strand SP, Johansen B (2016) The emerging patent landscape of CRISPR-Cas gene editing technology. Nat Biotechnol 34(10):1025–1031

Franco NT (2007) Un enfoque diferente para la protección de los conocimientos tradicionales de los pueblos indígenas. Estud Socio-Jurid, Bogotá, Colombia 9(1):96–129

ISF (2012) ISF view on Intellectual Property. Position Paper of the International Seed Federation, adopted in Rio de Janeiro, Brazil, 28 June 2012

Rapela MA (2005) Plantas transgénicas, bioseguridad y principio precautorio. Editorial de la Universidad Nacional de La Plata, 570 pages

Rapela MA (2006) Excepción del fitomejorador: de la libre disponibilidad a la variedad esencialmente derivada. In: "Innovación y Propiedad Intelectual en Mejoramiento Vegetal y Biotecnología Agrícola", Rapela, Miguel Ángel, (Director Académico), Gustavo J. Schötz (coordinador), Enrique del Acebo Ibáñez, Juan Miguel Massot, Helena María Noir, Fernando Sánchez, Andrés Sánchez Herrero, María Celina Strubbia y Mónica Witthaus. Editorial Heliasta, pp 207–242

Rapela MA (2008) El concepto de Variedad Esencialmente Derivada y la Excepción al Fitomejorador dentro del Derecho del Obtentor. 2° Congreso Nacional e Internacional de Agrobiotecnología, Propiedad Intelectual y Políticas Públicas. Universidad Nacional de Córdoba, 27 a 29 de agosto de 2008

UPOV (2004) Molecular Techniques. International Union for the Protection of New Varieties of Plants (UPOV) Administrative and Legal Committee. Fiftieth Session Geneva, October 18 and 19, 2004. Document prepared by the Office of the Union

WIPO (2015) Propiedad intelectual y recursos genéticos, conocimientos tradicionales y expresiones culturales tradicionales. http://www.wipo.int/edocs/pubdocs/es/tk/933/wipo_pub_933.pdf

Woese CR, Fox GE (1977) Phylogenetic structure of the prokaryotic domain: the primary kingdoms. Proc Natl Acad Sci U S A 74(11):5088–5090

Woese CR, Kandler O, Wheelis ML (1990) Towards a natural system of organisms: proposal for the domains Archaea, Bacteria, and Eucarya. Proc Natl Acad Sci U S A 87(12):4576–4579

Index

© Springer Nature Switzerland AG 2019
M. A. Rapela, *Fostering Innovation for Agriculture 4.0*,
https://doi.org/10.1007/978-3-030-32493-3

Printed in the United States
By Bookmasters